Advertisting a popular beverage, circa 1910.

SOFT DRINK BOTTLING

A HISTORY WITH SPECIAL REFERENCE TO ILLINOIS

John R. Paul

and

Paul W. Parmalee

Illinois State Museum Society
Springfield, Illinois
1973

CONTENTS

FIGURES

FIGURES—Continued

FIGURES—Continued

PREFACE

This book originated from a dialogue between the two authors. Although each of us had a limited facility with respect to bottle styles and their history, we appreciated the historic significance of the bottles and their manufacturers. The various questions that arose with respect to the age of the bottle, what it might have contained and the technology available at the time of its manufacture were seldom answered in readily obtainable books. A large number of our questions related to soft drink bottles. Difficulty in obtaining desired information on these motivated us to attempt to detail the history of this small segment of the bottle collecting field.

This undertaking may appear somewhat naive since the amount of information available on this single industry and its respective collectible items would fill many volumes. It is our purpose only to summarize the highlights so that the average collector is better able to evaluate his collection with respect to various historic points. While it is not possible to look into the history of every bottler in the United States (let alone in Illinois), the basic parameters and conditions under which these firms operated are set down within the text. We have selected, for matters of convenience, the bottling firms within the state of Illinois to use as examples whenever possible. There is every reason to feel that these examples are characteristic of bottlers throughout the United States.

The literature in this field has been very difficult to locate since few libraries contain material on the earliest days of the development of the bottling industry. We were forced to go to several major libraries scattered throughout the United States in an attempt to piece together the progressive development of the industry as reflected in its national journals. The *National Bottlers' Gazette*, the major trade journal of the industry, provided a great amount of information relative to the industry and the persons active within it.

Unlike several of the other books on soda water bottles, we have not attempted to show a random selection of bottles from a wide geographic area. The length of the table in Chapter X shows the great number of bottlers, i.e. bottle types and styles, from a single state. We hope that the basic format of our presentation helps the individual collector understand the enormity of his collecting field and offers guidance toward investigations of this type in his own respective geographic area. Our listing of Illinois firms is undoubtedly incomplete and only time will tell to what extent.

To the collector we offer our best wishes for success in an exciting venture into local history. We hope this book will form a foundation for many future studies.

John R. Paul
Paul W. Parmalee
Illinois State Museum

ACKNOWLEDGMENTS

All endeavors of this type can succeed only with the assistance of individuals who also share a personal interest in one or more aspects of the subject. Without the cooperation of these people, especially the bottle collectors and historians who have a firsthand knowledge of bottles, bottlers and companies in their city or area, this reference would fall far short of one its intended purposes. Many, such as Elmer Grubb of Peoria, Walter R. Sanders of Litchfield, James M. Sutherland of Joliet, Robert Kott of Summit and Richard Lipsky of Galesburg, have given a great deal of time to this project. We would like to express our sincere appreciation to them and to the following individuals for their assistance, great or small, in our effort to make this an informative and comprehensive volume. All are from Illinois unless otherwise noted.

George Bale, Auburn
Mrs. Floyd Barringer, Springfield
Mrs. Barbara Brown, Danville
John Bulthouse, Greenville
William Burge, Springfield
Mrs. Frank Carrithers, Toluca
Harry J. Coufal, Beardstown
Mrs. Maurice Exton, Roxana
Bernice Fernandes, Springfield
Ted Flowers, Rushville
Mrs. Marvin Ford, Jacksonville
Mrs. George E. Frank, Sheridan
Edwin C. Galbreath, Carbondale
David O. Geduhn, Cayuga (Indiana)
Roger D. Gustafson, Belvidere
Stan and Suzanne Hart, Oregon
Mrs. Robert T. Hastings, Pekin
Arlyn Heuiser, Collinsville
Bernice Hynd, Braidwood
Mrs. Harry E. Jahns, Sr., Batavia
Curt Johnson, Tiskilwa
Mrs. Jack Kirchner, Beardstown
Carl Landrum, Quincy
O. O. Lawson, Salem
Forrest D. Loomis, Monmouth
Charles J. Lorenz, Chicago
Ray Makkonen, Topeka (Kansas)
Jeanette K. Mattoon, Lake Forest
Harold F. Mayfield, Toledo (Ohio)
Mrs. McClain, Olney
Robert L. McNabb, Bondville
Arthur P. Montgomery, Richton Park
Wayne S. Nordstrom, Galva
Raymond E. Owen, Tuscola
J. David Parmalee, Springfield
Olga Pohlmann, Chicago
Terry Powell, Macomb
Mrs. Rosella B. Riley, Newton
Eugene Rittenhouse, Rushville
C. W. Sieber, Joliet
Dennis Stahlberg, Springfield
James E. Tanner, East Peoria
James Taylor, Chillicothe
Darwin Terwilliger, Wyoming
Edmund B. Thornton, Ottawa
Julian H. Toulouse, Hemet (California)
Hal G. Wagner, St. Louis (Missouri)
Miss Diane Wolf, Harrisburg
William R. Wolf, Mt. Carmel

We would like to express our appreciation to the directors and staff of the following libraries who so willingly made available to us volumes of the *National Bottlers' Gazette*: the New York Public Library, New York City, N.Y.; the John Crerar Public Library, Chicago, Ill.; and the Library, Graduate School of Business Administration, Harvard University, Cambridge, Mass. Mr. S. R. Kaplan, publisher of *Softdrinks* (formerly the *National Bottlers' Gazette*), graciously granted permission to incorporate in this work illustrations and other reference material from early issues of the *N.B.G.* We also acknowledge with thanks the administrators and staff of the Illinois State Historical Library and the Illinois State Archives, both of Springfield, for making city directories and manufacturing census data available to us; Mr. Thomas F. Baker, Executive Vice-president, National Soft Drink Association, Washington, D.C., who very kindly provided a copy of *A History of the American Soft Drink Industry* by John H. Riley, a reference which was indispensable in this study; and Thomas Nelson, Inc., Publishers, Camden, N.J., for generously granting permission to utilize certain specific data presented by Julian H. Toulouse in his recent volume, *Bottle Makers and Their Marks* (1971).

A special expression of gratitude is extended to the following individuals for their valuable help with this project: Marlin Roos, Museum photographer, for the difficult and detailed work involving most of the figures; Lynda Thompson for assistance with the organization and layout of the figures; Violet Jones for her efficient work in typing the manuscript and in numerous other aspects involved in its preparation; and Orvetta Robinson, Museum Librarian, for critically editing the manuscript and handling the varied tasks relative to its publication.

Lastly, we are deeply indebted to and most appreciative of the various antique dealers, serious bottle collectors and all others who patiently permitted us to record information from bottles and other related materials in their collections.

SOFT DRINK BOTTLING

A HISTORY WITH SPECIAL REFERENCE TO ILLINOIS

Chapter I
THE EARLY HISTORY OF MINERAL AND SODA WATERS

Soda water as we know it today had a rather inauspicious origin in the early apothecary shops of 18th century England. In the latter half of that century, Thomas Henry is credited with the commercial manufacture of a tonic known as "Mephitic Julep." In these early days "mephitic" was a term typically applied to carbon dioxide although today it means "noxious" and is fittingly the scientific name of the striped skunk. This initial attempt to sell a carbonated beverage was accomplished by promoting it as a form of medicine said to be helpful in cases of "putrid fevers, scurvy, dysentery and bilious vomitings" preferably taken with "a draft of lemonade." Thus, one of the first flavored carbonated beverages was formed. The name "soda water" was applied to such beverages around the beginning of the 19th century; and the term "soda pop" appears to have been accepted as the standard term for soft drinks by the time the Hutchinson-type stopper was developed about 1880.

Although these earliest manufacturing attempts were primarily of English or European origin, the American scientists of the day were not long in appreciating their importance. Benjamin Silliman of New Haven, Connecticut, was one of the earliest manufacturers (circa 1806). Silliman's problems in the industry were summed up in a letter of complaint sent out by him in search of suitable bottles: "I cannot procure any glass bottles which will not burst nor any stone ones which are impervious to the fixed air." Pottery bottles known to have been used for soft drinks are not commonly encountered (see Fig. 1).

These early efforts to commercialize the manufacture of soda water were not oriented to the luxury drink trade but were directed toward the medicinal market. Thus, the highly carbonated soda water fitted well into the pharmacopeia of the day. Although soda water evolved rather rapidly from a medicinal beverage to a flavored refreshment in the American market, in England it was considered a medicinal product until 1898 and, as such, contained sodium bicarbonate according to pharmacological requirements.

In the first pharmacopeia of the United States (1820), soda water is described as containing two ounces of subcarbonate (bicarbonate) of soda to one gallon of water and impregnated with carbonic acid gas equal to about ten times its volume. No further reference was made to soda water in any United States pharmacopeia although carbonic acid water (water impregnated with carbon dioxide with soda added) was listed until 1870. Carbonic acid water is simply plain carbonated water and is basically identical to the product to which the name "soda water" is currently applied. As early as 1820 this same combination was also referred to as "seltzer water." After this initial description in 1820, the level of carbonation was reduced to five times the bulk of the water, approximately the same as that used in carbonated beverages today. The use of soda water in the United States as a medicinal product significantly declined between 1830 and 1850 although soda water and mineral waters were still widely acclaimed as medicinally valuable until well after the turn of the 20th century.

Figure 1. Mid-19th century pottery soda bottle.

Only the addition of flavoring and a sweetening agent was necessary to bring about soda water as we know it today. It appears that, soon after the commercial development of the unsweetened product, various experimenters began adding sugar or other substances to impart a better flavor to the rather flat "mephitic" drink. Citrus juices had long been known for their medicinal and flavoring qualities. Lemon, more than any other, was used to alter the flavor of these early drinks. Other persons experimented with wines, adding some carbonated water to both lighten the wine and make it last longer. As early as 1809, reference was made to adding both wine and sugar to soda water to make a refreshing and "grateful" drink.

A forerunner of modern instant drinks, powder for the preparation of mineral water and soda water was available from the earliest days. One firm in business in the early 1800s advertised "carbonated sodiac powders for making soda water" with instructions to "dissolve the powders contained in the white and blue papers in separate tumblers with about one gill of water each; mix and drink immediately." Other products known to have been sold included "Farr's Soda Powder" for making soda water;

FIGURE 2. Ready-mix mineral salts for home manufacture of mineral waters.

"Milnor's Acid Lemon" for making lemonade; and "Carpenter's Saratoga Powders" for making Congress Springs or Saratoga waters, long a favorite medicinal water. Although bottling of the artificial soda and mineral waters had been done since the beginning of the 19th century, it was well into the 1860s and 1870s before sales of the bottled product became commonplace.

In a catalog published by James W. Tufts about 1885, ten different types of prepackaged dry contents for the manufacture of mineral waters were offered. Each one-pound package was sufficient to make 10 gallons of potable beverage, including such standards as Vichy, Congress, Excelsior, Apollinaris, and Saratoga. These products were advertised as essentially identical in taste and chemical properties to the natural mineral waters whose names they carried and were said to contain the same medicinal qualities. Many other firms produced similar packaged mineral water salts (Fig. 2).

Although the medicinal benefits of mineral waters and carbonated beverages were not definitely established, such claims attached to them did not diminish until well into the twentieth century. One of the most widely distributed of all commercially sold mineral waters was "Pluto" (Fig. 3). Advertised as "America's Physic," it contained a spring water fortified with sodium and magnesium sulfates. This product, bottled in French Lick, Indiana, from the late 19th century until well into the 20th century, was sold over a wide geographic area. As in the case of most other mineral waters, it was largely touted as a laxative and purgative.

One Illinois firm that deserves special mention here, because of its varied products and the general business operation typical of the period, is the Gravel Springs Company of Jacksonville. Among its many products was the Gravel Springs lithia water, "Hy-Rock Lithia," a natural mineral water derived by mixing 20 grains of lithia with each gallon of spring water. Another product of this firm was "Tritti-Carbo," a derivative of the common couch grass, which was considered beneficial in treatment of "all inflamed and congested conditions, whether of the stomach, bowels, or urinary passages," and reputed to exert a decided influence on diseases of the prostate due to old age. Perhaps the most potent of the Gravel Springs beverages was "Nix-E-Flyer," said to cure "headache, nausea and that peculiar 'how-came-you-so' feeling occasioned by over-exertion, biliousness, dyspepsia, or the too strenuous pursuit of pleasure." Nix-E-Flyer was considered one of the very best laxative waters, "invaluable for the relief of all torpid conditions of the liver and alimentary tract," and its advertising slogan was "Through Without a Stop."

FIGURE 3. Pluto Water, one of the better-known nationally distributed brands at the turn of the century.

Chapter II
THE SODA FOUNTAIN

Relatively few American businesses were known to have actually bottled mineral and spring waters for consumption away from the premises during the earliest years of the trade. It was not until improvements in bottle and closure production were made that significant numbers of bottled waters became available for widespread public consumption. Thus, in the early years the manufacture of these beverages was largely confined to the soda fountain.

With an increase in industrialization in the early 1800s, many inventive efforts were directed toward the development of manufacturing equipment capable of producing and dispensing carbonated beverages. The first known patent for such a procedure was issued to Joseph Hawkins in 1809 for preparing artificial mineral waters. In 1810 the firm of Simmons and Rundell (according to the *National Bottlers' Gazette*) of Charleston, South Carolina, (spelled *Simons and Riondel* in Riley, 1958) was granted a patent for the technique of saturating water with fixed air (CO_2). These early patents were coincidental with the production of soda and mineral waters from fountain apparatus by Silliman of New Haven and various firms in New York and Philadelphia. These early fountain dispensers, soon to be taken over by the apothecary shop for a period of at least 20 years, apparently served a substantial number of people during their brief period of use. The tendency of patrons to drink the waters for their medicinal properties rather than refreshment value hastened the demise of the initial fountains except when they were operated as part of the pharmaceutical business.

Although no detailed descriptions exist of the earliest soda fountains, the one developed and manufactured by John Matthews about 1830 was basically the style used throughout their history. Initially, the apparatus consisted of a tank of carbonated liquid encased in ice beneath a counter; from the tank a tube projected above the counter through which the pressurized liquid was forced by the expanding properties of the compressed carbon dioxide. Such a tank was, from its conception, referred to as a "fountain." Other minor adaptations of this basic design were quickly effected; however, none substantially modified the principle involved. The modern concept of a fountain as the table or counter where sodas or ice cream are served is a derivative of the original fountain which was actually the container under the counter from which soda water was dispensed. The dispensing fountain is still located beneath the serving counter in most modern fixtures.

Carbonation was accomplished by one of two possible methods. One procedure involved mixing an acidulant, for example, tartaric acid or sulfuric acid, with an alkali, such as super-carbonate of soda, marble dust or their equivalents, in the container of water to be carbonated. The other method entailed the creation of carbonation in a separate vessel; the resulting CO_2 was then piped into the container of water which was agitated until it became thoroughly saturated.

Portable fountains were constructed of copper with a tin-washed interior surface. Earlier uncoated fountains were found to be of considerable danger to the consumer since the acids used in carbonating reacted with the copper and formed compounds which resulted in occasional cases of nausea and poisoning. Lining the interior of the fountain with a thin coat of molten tin prevented such poisoning; however, constant use eroded the internal lining and eventually exposed areas of raw copper. Retinning the surface maintained the useful life of the fountain; but since the internal portions of the fountain were not easily observed, it was frequently only after a customer became ill from drinking the soda water that erosion of the tinned surface was discovered.

Production of cast-iron fountains, which presented fewer problems from a corrosion standpoint and far fewer difficulties from structural weakness, began in the 1840s. During these early years, however, the extreme weight of the iron fountains hindered their general acceptance by the entire market. The first iron fountains were lined with paraffin, porcelain or glass, later ones with sheet block-tin. Paraffin linings were cheap but wore out quickly; porcelain linings cracked and chipped; and glass linings, although chemically neutral and thus perfect for the purpose, were fragile and did not hold up under sustained use. Thus the sheet block-tin linings proved to be the best and most efficient for continued use. One of the greatest difficulties in the manufacture of these fountains was developing sufficient strength to keep them from exploding when pressurized. The fountains were typically charged with 150 pounds per square inch of carbonation, and only those well built held up through continued use under these conditions.

These fountains were designed to be removed from the dispensing device and returned to the carbonator who would recharge and return them to the dispensing locality. The basic technique for charging a fountain involved filling the device 2/3 to 3/4 full of water and then admitting the gas until a pressure of approximately 150 pounds was attained. The two factors that controlled the relative amount of gas absorbed by the water were temperature and pressure. The lower the water temperature, the greater the volume of gas which would be dissolved; the higher the pressure of the gas, the greater the amount of gas dissolved. Because most of the gas rushes out of the solution when the pressure is reduced to atmospheric levels, it is largely only that gas contained in the solution, because of its low temperature, that remains in the consumer's drink. Thus, the relatively high pressure to which fountains are charged is convenient for emptying the fountain

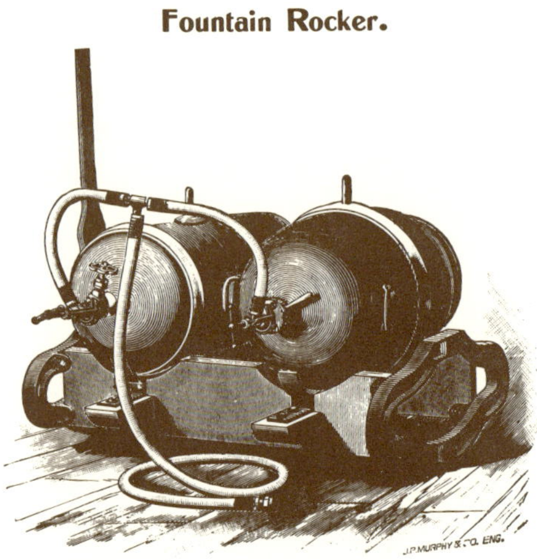

FIGURE 4. A fountain rocker used to mix CO_2 and water (Tufts Catalog, ca. 1895).

rather than for maintaining a high pressure within the carbonated beverage. However, to assure a thorough mixture of the gas and liquid, mechanical agitation of the fountain on a rocker device (Fig. 4) was necessary in order to create the highest possible carbonation.

Although some drinks were formed by drawing carbonated water into a glass containing flavoring and ice, many beverages such as root beer, ginger ale and sarsaparilla were charged in portable fountains. In such cases, approximately one gallon of syrup was added to nine gallons of water and charged to a pressure of approximately 90 pounds rather than the normally higher charge for plain soda.

A description of Delatour's Soda Water Stand on Wall Street in New York City illustrates the manner in which many concerns distributed soft drinks in the 1870s.

> The fountains were placed in an ice house and thoroughly refrigerated before charging — the cold water, of course, imbibing the largest amount of gas. Then they were snuggly packed with ice in a large tub, as ice cream freezers are packed and conveyed to the Wall Street store. Reaching there, a "goose-neck" was attached to the fountain and without removing it from the ice its contents were drawn for the customers, scores of whom often stood in a long queue waiting their turns. The syrups used were extraordinarily thick and rich and were kept in bottles on ice until they were called for. The trade was limited to the four or five hours of greatest activity in Wall Street, but the business done has, perhaps, never been surpassed, or even equalled, at any other soda water stand in the world.

Needless to say, many problems evolved in using the portable fountains. The cleansing of the fountain between uses was a point of contention, and the better distributing houses frequently bragged that their cleaning, if nothing else, made their product superior to their competitors. No one drinking a glass of plain soda water from the bowels of the fountain wanted to detect even a hint of sarsaparilla or root beer flavor!

Other problems constantly faced the fountain vendor. The sheet block-tin linings were susceptible to collapsing if internal pressures subsequently became less than the external atmospheric pressure. Although this was not a major problem, it happened often enough to make it of considerable concern to the vendor. Likewise, the iron and steel fountains corroded rather rapidly and required constant preventative maintenance and painting. The copper fountains, although dangerous in other respects, could be kept in good condition by an occasional scouring and cleaning of the external surfaces. Explosions of the pressurized fountains, although not common, also occurred with sufficient frequency to warrant an active program of inspection and maintenance. Relatively lax inspection and testing procedures in the manufacture of fountains occasionally made them less than perfect for use. As F. E. Saunders noted in the *National Bottlers' Gazette* in 1891, "A careless bottler may work his horse until it dies in the harness and incur no risk except a small fine from the S.P.C.A., but no one can continue to use a dilapidated fountain without assuming a grave responsibility and risk."

Fountain distributors faced a problem similar to that of a soda pop bottler, that is, the return of his container. In the case of bottles, apparently no system of return deposits had been devised in these years of the 1800s. It would have been considerably more difficult to require a deposit on the fountains since the package was worth twenty-five to fifty dollars and the contents only one to two dollars. Apparently a large number of fountain distributors were sufficiently dishonest to abscond with a competitor's fountain which they could then refill and sell as their own. With such dishonesty present, the painting of a name or number on the fountain offered little protection to the distributor.

Recommended procedures of the day included having the name of the fountain owner attached to the fountain in brass letters by the manufacturer. However, it was warned that the addition of such identifying letters or numerals after the fountain was completely assembled was injurious to its proper function. The heat necessary for application of the identification code would damage the lining and spell quick doom for the device. And, alas, in the days without computers, it was difficult for some of the less businesslike firms to keep track of the customers using their fountain devices. When the considerable expense of the fountain was compared to the product inside, the worried distributor could ill afford to have an illiterate driver fail to keep track of the fountains he delivered. The trade literature offers numerous devices for developing card files, delivery books, and other simple procedures to maintain a constant surveillance on the manufactured product. Empties had to be washed, refilled and returned to the dispensers with haste, and much effort was expended to see that they were not lost enroute.

Arguments over the merits of soda fountains and various related devices, not to mention suggested improvements and actual patent rights, ensued for the next 50 years. The ad placed by the Iron Clad Manufacturing Company, stressing their victory over the Matthews firm in a court battle over patent infringements, seems to be typical of these 19th century competitors. This court decision in August of 1884 would suggest that the Iron Clad Manufacturing Company and the Matthews firm were at each other's throat (Fig. 5). However, in 1891 the four largest firms engaged in the manufacture of soda water apparatus consolidated and organized the America Soda Fountain Company. These four firms were James W. Tufts, A. D. Puffer & Sons, Charles Lippincott & Com-

THE IRON CLAD MAN'F'G CO. VICTORIOUS!

MATTHEWS' PATENT DECLARED VOID.

The Iron Clad [Steel] Fountain no infringement and the only Safe Fountain Made.

AS IT WAS THEN.—From "The Bottlers' Gazette," October, 1883.

TO THE BOTTLING TRADE.—The long contested suit of the firm of **JOHN MATTHEWS vs. THE IRON CLAD MANUFACTURING COMPANY**, in the matter of the patent steel riveted fountain, has been decided, and Judge Blatchford, of the U. S. Circuit Court, has ruled that the Iron Clad (Steel) Fountain is not only not an infringement, and placed the costs of the suit upon the firm of John Matthews, but has also decreed that the Matthews fountain patent is null and void. According to this decision anybody can make and sell the so-called steel fountain, as made by Matthews, and leaves the patent riveted truss bottom steel fountain the best and only safe fountain in the market. We congratulate the trade on this decision, for it leaves everyone free to use his own judgment. Thanking you for the interest shown, we are, joyfully,

IRON CLAD MANUFACTURING CO.,
Manufacturers of the Iron Clad Steel Fountain,
22 Cliff Street, New York.

WHO KILLED THE MATTHEWS FOUNTAIN?

1. WHO KILLED THE MATTHEWS FOUNTAIN?
I, SAID THE IRON CLAD,
AND EVERYBODY'S DEVILISH GLAD,
I KILLED THE MATTHEWS FOUNTAIN.

4. WHO MADE ITS SHROUD?
WE, SAID THE CROWD
OF MACHINERY MEN, PROUD,
AND WE MADE ITS SHROUD.

7. WHO'LL CARRY IT TO THE GRAVE?
WE, SAID THE "BOYS,"
WITH A SHOUT AND GREAT NOISE,
AND WE'LL CARRY IT TO THE GRAVE.

AS IT IS NOW—Judge Blatchford's Decision, U. S. Circuit Court, August 20th, 1884.

2. WHO SAW IT DIE?
I, SAID DeLACY PETE,
I WRAPPED ITS WINDING SHEET,
AND I SAW IT DIE.

5. WHO DUG ITS GRAVE?
THE FIRM'S OWN GALL,
WHICH THEY SPREAD BEFORE ALL.
THAT DUG ITS GRAVE.

8. WHO'LL BE CHIEF MOURNERS?
WE, SAID CONNER AND WHITE,
FOR THIS ENDS THE FIGHT,
AND WE'LL BE CHIEF MOURNERS.

3. WHO CAUGHT ITS BLOOD?
WE, SAID THE PUFFERS,
WE'RE THE MERRY DUFFERS,
AND WE CAUGHT ITS BLOOD.

6. WHO'LL BE THE PARSON?
I, SAID TUFTS, BRAVE,
THAT LONOR I CRAVE,
AND I'LL BE THE PARSON.

9. WHO'LL SING THE PSALM?
WE, ALL THE BOTTLERS SAID,
FOR YEARS ON US THE FIRM HAS FED,
AND WE'LL SING THE PSALM.

FIGURE 5. Patent fights were commonplace and boisterous during the late 1800s. The battle over the patent steel-riveted fountain was widely expounded in industrial literature (*N.B.G.* 1884).

FIGURE 6. "The Celtic," a typical soda-counter dispenser used by the small retailer (Tufts, 1895?).

pany, and the John Matthews Apparatus Company. Within a few years this consolidated firm also purchased the stock of the Hartt Manufacturing Company of Chicago, one of the major midwestern manufacturers of soda water apparatus.

A new trend in fountain dispenser design, which in later years would become more nearly the standard, resulted from the development of the marble soda fountain by the A. W. Dows Company. In 1861 this firm patented a fountain which consisted of a marble box containing a cooling coil for soda water, syrup containers, and an ice shaver. Shaved ice was placed in the drinking glass, syrup added, and then soda water poured over the top. Ornate faucets and decorative features gave the dispensing apparatus a distinctive appearance. This marble device was exhibited in Paris at the Great National Exhibition in 1877 where it was awarded medals and high commendation. In later years, several firms including John W. Tufts of Boston (Fig. 6), manufactured and sold these marble "fountains."

A more simplified carbonating system was developed by Scannell of Boston in 1884. In the Scannell device, gas and water were pumped simultaneously through a single pipe into a narrow vertical mixing chamber where the water and gas were forced through a series of perforated disks in the form of a spray. The water in this fashion attained a high degree of carbonization without the use of separate agitators, thus avoiding the use of labor or power upon which the older mechanical methods depended. In the years to follow, several other firms developed methods that were similar to this and which used the same principles of operation. The ultimate method of carbonating beverages has subsequently been the use of liquid CO_2.

Although early European experimentation with liquefaction compressors yielded small but functional amounts of the liquid gas, it was not until 1879 that a German scientist succeeded in producing a sufficient quantity for commercial purposes. In that year the usability of this system for raising sunken objects from the ocean floor was demonstrated. A deflated balloon was attached to a ship's anchor resting on the bottom of Kiel Harbor. This balloon, subsequently inflated with liquified CO_2, quickly raised the anchor to the surface. With its potential uses thus shown, developers soon promoted liquified CO_2 for pressurizing beer containers and making soda water. By 1883, a mineral water bottling plant in Berlin, Germany, was successfully employing liquid CO_2 in its business. In 1886 some attempts were made by this firm to sell the liquid gas to American bottling firms for commercial purposes, but difficulties arose and the use of the liquid gas by American companies was not immediate.

One of the earliest American producers of liquid CO_2 was Jacob Baur of Chicago who organized the Liquid Carbonic Acid Manufacturing Company in November, 1888. This firm developed the system by which the gas was derived from combustion, compressed and liquified, and then permanently stored in steel cylinders. Other firms were soon to enter the competitive market, and by the mid-1890s seven or eight firms were known to be manufacturing the product in a similar manner.

Two factors influenced the changeover from mechanical carbonators to liquid CO_2. First, the cost of the liquid gas at the time of its introduction to the trade in 1886 was approximately twenty-five cents a pound, more than three times the usual cost to most bottlers to make the product in their own plants. Second, virtually all bottlers had the equipment to manufacture CO_2 and were not in any hurry to dispose of it as long as it was usable. However, employment of liquid CO_2 soon increased and, by 1906, 15 million pounds were produced annually (Riley, 1958). By this time the market price had dropped to eighteen cents a pound and the carbonic acid industry was well established.

Perhaps the greatest advantage of this means of carbonization was the elimination of large quantities of machinery and labor that were necessary in the manufacture of the gas. The danger of handling the sulfuric acid and the occasional explosions from overpressurization all lent their weight in favor of adoption of the relatively safe prepackaged product. Not only did the introduction of liquid CO_2 simplify the manufacture of a uniform quality in the carbonated beverage, but it also eliminated much of the prejudice against the industry's products that were based on a beverage "made from marble dust and sulfuric acid."

The soda fountain perhaps reached its highest pinnacle of success during the Philadelphia Centennial Exposition of 1876. James W. Tufts, joined by the Charles Lippincott Company, paid $50,000 for exclusive rights to advertise the soda fountain and serve soda water at the Exposition. Their main exhibit was a soda fountain over 30 feet in height which, in addition to performing normal duties of the fountain, displayed rare exotics, ferns, and flowers, and was topped by a large illuminated dome. The exterior apparatus was silver-plated and liberally decorated with statues, globes, and other items to attract the eye. In addition to this large fountain, several smaller fountains were located throughout the grounds. Although the amount of business done was reported to have been enormous, it was not a profitable undertaking in itself; but it did prove an invaluable advertisement for the fountain (Riley, 1958: 66-67).

The commercial manufacture of carbon dioxide in a liquid form that could be sold to druggists eliminated the need for the individual soda fountain operator to generate his own gas. However, it did not eliminate the use of this older method, and generating CO_2 at the soda fountain continued for many years thereafter. But alas, the ornate marble structure of the mid-Victorian soda fountain, which was frequently positioned as an island in the midst of the drugstore, apothecary or candy-sellers displays, was superceded by the type of fountain which was built along one wall of the store.

The Syphon Bottle

A rather small development in the early soda water industry was the use of the syphon bottle. Essentially designed as a miniature precharged soda fountain, this device allowed a portion of the bottle's contents to be used while retaining the carbonation in the portion remaining in the bottle. The earliest patent for this type of bottle, in 1813, used the principle currently followed although it functioned with a stopcock rather than a spring-operated valve in the stem. In 1829 two Frenchmen patented a hollow corkscrew which was inserted through the cork into the charged bottle of soda water, the spout portion of the corkscrew containing a spring-operated valve. Thus, the liquid was able to escape the bottle only through release of the valve. By the late 1830s, syphon bottles of the same basic design as those in use at the turn of the 20th century became a standard device. The early use of these bottles was restricted primarily to the European market. Although the principle involved was utilized in the soda fountains of the mid-1800s in America, it was not until later in the century that the actual home use and consumption of such unflavored soda waters and carbonated mineral waters became popular in the United States. James W. Tufts (1885?) emphasized the value of the syphon bottle especially to the saloon keeper and to patients for whom physicians prescribed plain carbonated waters for home use.

One of the primary dangers in the use of this syphon bottle was the extreme pressure under which it was filled. Tufts notes that all syphons are charged to 150 pounds per square inch or over. The syphon fillers, illustrated in Figure 7, all exhibit a sturdily constructed guard into which the bottle was inserted for the filling procedure. Only the best and strongest of glass containers withstood the pressures of filling and the days of hard use by the consumer. Several turn-of-the-century firms such as "The New York Bottlers' Supply Mfg. Co." sold quite a varied line of syphons and syphon heads (Fig. 7), many of which were manufactured in Europe.

Directions for filling the syphon bottles were always explicit in one area: the bottle must not be filled more than 4/5 full to allow sufficient space for gas enough to empty the contents. Another and rather unusual advantage of the syphon bottle was its use as a fire extinguisher. Being filled with carbonic acid gas and water, both of which were noncombustible, the syphon was promoted as a suitable substitute for the other fire extinguishers of the day. The drinking public had only to keep on hand several full syphons of this delectable product for use in quenching a thirst or a lively blaze.

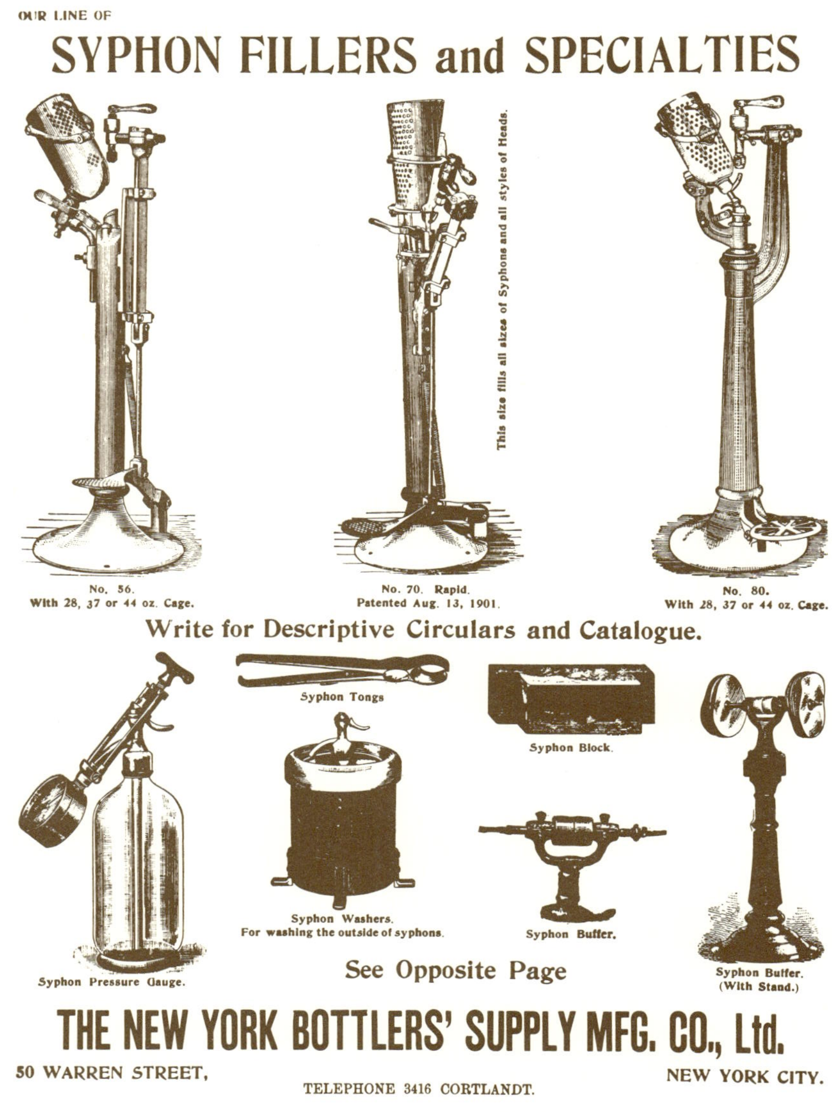

FIGURE 7. Typical advertisement for syphons and accessories (*N.B.G.* 1901).

FIGURE 8. Examples of syphon bottles used by Illinois firms.

Syphon bottles, perhaps more so than bottles used for soda pop, were frequently found in varying shades of colored glass with blues and greens predominating. Most of the earlier syphons had the name of the bottler etched lightly into the surface of the glass rather than embossed. In the 1930s, as with the soda pop bottles, the enamel label was painted onto the surface of the bottle. Early Illinois bottling firms, such as Blum's of Galena, H. W. Dresch of Springfield, Sanicula Springs of Ottawa, A. D. Huesing of Rock Island and Enterprise Bottling Works of LaSalle (Fig. 8), to mention only a few, employed and serviced these syphon bottles as an important sideline in their regular soda pop bottling operation.

Chapter III
CLOSURES

The search for an efficient and economical closure for the soda pop bottle was a furious and highly competitive battle. Perhaps only barbed wire, with its innumerable types developed and patented in the late 1800s and early 1900s, went through as much competitive and varied development as did the bottle closure. Literally thousands of patents were issued for various types of bottle closures between 1850 and 1900 (Fig. 9). Despite the many closures which were developed, virtually all fell into less than half a dozen basic types or categories. Even among those patents which gained acceptance within the industry, relatively few ever dominated the field as did the original cork stopper, the Hutchinson stopper of the 1880s and 1890s, and the crown stopper which is the type used today.

Corks

The earliest stoppers were made of cork and held in place by various sealing compounds (e.g., wax), wire bales, and the like. Over the years other types of stoppers were devised, but cork still continues to play an important role in the bottling industry. Currently, many wines and champagnes are sealed with cork stoppers in a manner reminiscent of the early 19th century. However, one of the major disadvantages of the cork stopper is its tendency to shrink when dry. This was of great concern to the carbonated beverage manufacturer since any leaks in or around the closure device might result in not only loss of carbonation but also seepage of the contents. It was difficulties of this type that first precipitated the development of the round-bottomed or pointed-bottomed bottle (Fig. 10). By using a bottle that could not remain in an upright position by itself, the bottler was assured that his product would be stored in a horizontal position. This kept the liquid in the bottle in direct contact with the cork, thus assuring that the necessary moisture would be retained in the body of the cork to keep it swollen and tightly sealed. For the purpose of temporary display or as a rest for these bottles while the contents were being consumed, special holders (Fig. 11) were available. It now seems rather unusual that round-bottomed bottles were produced possibly as late as 1905 to 1910 to accommodate the crown cap (Fig. 11). Keeping the cork wet to insure a tight seal evidently was still a worry. Some flat-bottomed crown soda bottles of still later vintage possessed paper labels stating "Lay this bottle on its side."

Corks were inserted by hand in virtually all bottles and, therefore, did not require extensive capital investments in expensive sealing or capping machinery, a factor which has been a consideration for retaining cork as a bottle seal since these early days. It was not long, however, before some fairly simple wiring devices were developed (Fig. 12) which allowed the large-scale bottler to seal—and keep sealed—the rambunctious beverage. An interesting 1890s Hires ad illustrated a technique for resealing the soda pop bottle at home by hand-tying a string or wire retainer in the recommended fashion.

The *National Bottlers' Gazette* was quick to point out that the use of cork stoppers led many people into lives of avarice, greed and ill repute, implying that there was no greater crime in the late 1800s than pulling the cork from a champagne bottle, "cleaning" it off in an "unsightly" fashion, recutting it, and inserting it in another bottle at a later time. With payment as high as sixty cents a pail for clean champagne corks, and thirty-five to forty cents a pail for those from ginger ale, soda water, sarsaparilla and other "promiscuous" bottles, one could literally double his income with this illicit business.

Internal Stoppers

Early efforts to find a replacement for the cork included experiments with other types of internal stoppers. Although never popular in the United States, internal stoppers other than cork have been used to a considerable extent in many European and Asian countries. One of the earliest American patents of this type, issued to John Matthews in 1864, was known as the Matthews Gravitating Stopper. This device was an internal metal stopper shaped in the form of an elongated plug which was pulled into the closed position by a magnet. Its advantage over other types of internal stoppers was reputed to have been its ability to withstand constant use without becoming deformed and thereby causing a break in the seal. However, one of the major disadvantages of the Matthews stopper was its protruding stem which could easily be damaged or dislodged accidentally, thus causing a loss of the entire contents of the bottle.

An early Illinois soda bottle (Wm. Bower, Olney; Fig. 13, c) has been found that is thus far unique among all similar styles. The bottom is embosed GRAVITATING STOPPER/MADE BY/JOHN MATTHEWS/NEW YORK/PAT OCT 11, 1864 (Fig. 13, a). It is reasonable to assume this bottler ordered his bottles from the Matthews firm and that Matthews used this embossing to advertise his stopper. Many of these bottles were excavated near Newton, Illinois, and several still possessed portions of the metal collar and hinged flap (Fig. 13, b) that served to hold the stopper in place.

Other types of internal stoppers patented on or about that time included the "Codd" with its internal glass stopper (Fig. 14) and several other varieties which featured hard rubber or metal sealing balls. These latter types all operated on the same basic premise: (1) the bottle was first filled with the highly carbonated beverage; (2) the bottle was then inverted, the heavy ball stopper settling against a gasket in the neck of the bottle; and (3) the force of the carbonation held the stopper in place

FIGURE 9. Some of the bottle stoppers awarded patents for 1885 (*N.B.G.* 1886).

FIGURE 10. Round- and pointed-bottomed bottles typical of the mid- and late 19th century.

after the bottle was returned to the upright position. When the consumer wished to drink from a bottle of this type, he needed only to force a narrow object into the neck of the bottle to dislodge the ball from the sealing ring which permitted the carbonation to escape; the ball then fell to the bottom of the bottle. The major disadvantage of this system was the absolute impossibility of resealing the bottle without losing the contents. Although the Codd style of bottle was common in England and many other parts of the world and widely advertised in the United States, apparently it was seldom if ever utilized in large operations in this country.

A minor variation of this internal seal involved a ball that floated (Fig. 15). In this instance a hollow steel or rubber ball was used although the actual sealing procedure was essentially the same. The frothy foam from the carbonated beverage was sufficient to hold the floating seal in place without turning the bottle upside down to permanently set the ball. Internal ball stoppers had one advantage over protruding types of seals in that they could rarely be accidentally dislodged.

HUTCHINSON-TYPE STOPPERS

Spring-loaded internal stoppers were patented in great numbers, but it was the one designed and sold by the W. H. Hutchinson & Son Company of Chicago which came to dominate the field. Numerous patent battles were fought over this type of stopper, with the subsequent yielding of most competitors to the original Hutchinson seal. This closure consisted of a wire loop attached at one end to a rubber disk which fitted snugly against the inside shoulder of the bottle. This flexible wire held the disk by spring tension either in the sealed position or, when depressed, in an open position to allow passage of the contents. The various spring-wire designs included oval, figure eight and other similar configurations. Perhaps the one feature which brought about the absolute popularity of the Hutchinson stopper was its ability to be closed by the consumer, thus allowing a portion of the contents to be saved for later use. The Hutchinson bottle could be reused again and again, employing the same stopper mechanism. The stoppers were economical, selling for $2.00 to $2.50 per gross (Fig. 16). It is reputed that by 1890, about 10 years after the patent date, more than three thousand customers were using the Hutchinson-style stopper. Most of the users of this closure device were soda water bottlers, and this stopper is considered a characteristic soda water seal.

Patent infringement or alleged infringements were apparently common in the latter 1800s, not the least of which centered around the famous Hutchinson stopper. Considerable verbiage was expounded by W. H. Hutchinson & Son and by Edward H. Everett, a glass and closure manufacturer of Newark, Ohio, relative to whose patent was

FIGURE 11. Holder for the round-bottomed bottle advertised by Tufts (1895?), and a crown round-bottomed bottle of the 1905-1910 era.

FIGURE 12. The Lightning Bottle-Wiring Machine did "the work of five men" (*N.B.G.* 1886).

actually the first. Figure 17, from the *National Bottlers' Gazette,* gives both sides of the argument, although probably only the original participants had any true knowledge of the actual dates their respective products were developed. However, the United States Circuit Court ruled in favor of the Hutchinson firm, thus giving sufficient impetus to this well-liked supplier to make his seal the dominant closure from 1880 to 1905.

Stoppers of the Hutchinson type were used and reused until well into the 1900s. The manufacture of this type of closure was discontinued by Hutchinson in 1912, but it was continued by some of his competitors until a much later date.

(c)

(a)

(b)

FIGURE 13. An early soda bottle from Olney, Illinois, illustrates an embossed ad on the bottom for the Matthews Gravitating Stopper (a & c) and the collar (b) used to hold the stopper in place.

FIGURE 14. Two examples of the Codd-style bottle that utilized glass marbles as stoppers.

Swing Stoppers

Coincidental with the development of the Hutchinson stopper was the manufacture of the "swing" stopper by Henry W. Putnam (Fig. 18). Although suitable for sealing all types of carbonated beverages, this closure was quickly adopted to seal beer and ale bottles. Apparently it was seldom used on any but the largest of soda bottles.

The swing stopper consisted of a hard rubber plug or disk which fitted over and partially into the top of the bottle, and by means of wire loops was permanently attached to the bottle neck. Wire fulcrums of various designs were attached to the closure; considerable leverage was needed when snapping the closure into place. This feature allowed an air-tight seal to be applied to virtually any bottle top. As with many of the other types of stoppers, literally hundreds of similar patents and improvements on the Putnam design appeared in the next 10 to 20 years. Why certain stoppers became popular and other similar styles failed is unknown; however, the name and reputation of the closures produced by the Putnam firm were apparently sufficient to allow the "Lightning" bottle stopper to dominate the field. Similar stoppers of this type, which used a porcelain seal surrounded by a rubber gasket, were subsequently developed. In effect, the device duplicated the results of the all-rubber Lightning stopper but was perhaps less susceptible to destruction resulting from continued use. Should the rubber gasket on the porcelain stopper become worn, a replacement could easily be installed, reconditioning the device to near-new condition.

Other Stoppers

Among the other closure devices to be merchandised in the United States were the internal screw-stoppered bottles (Fig. 15). In this style, a hard rubber or cork stopper was manufactured with threads which matched those of the inside of the bottle neck. This type of stopper, although widely used by English firms for beers and mineral waters, was seldom employed by American firms. Although both the internal screw stopper and the Codd (marble-in-neck) bottles were essentially restricted to England and the Continent in use, they were offered to the American market in most of the common trade journals. Although we have not found that any United States firms definitely used these styles of bottles, undoubtedly several did. Illustrations from the *Old Bottle Magazine*, July 1969, show both the Codd and internal screw stopper on the cover, and Muncie (1970) makes note of the same types of bottles in his book. However, all bottles illustrated in both these sources were from firms in Hawaii. The Hawaiian Islands, a United States protectorate since about 1900, were initially influenced by English missionaries and explorers. The predominate cultural influence during the period from the 1860s to the 1880s was essentially of English origin, and this factor probably explains the use of these two styles of bottles in the Islands.

Another type of seal, called the "New Bottle Seal" (Fig. 19), utilized a small disk of cork or rubber placed inside the neck of the bottle which remained in place by its own tendency to expand. By spearing these seals with an ice pick or similar device, one could quickly remove the seal and empty the contents. A device for resealing this bottle (Fig. 19) utilized the cork-like shape of the bottle opener as a self-sealing device.

Another use of wire ties for the sealing of cork stoppers involved the development of the "Bernardin Bottle Cap" (Fig. 20). This closure was essentially a complete sheet-metal cap which fitted over the top of the cork and under the finished crown of the bottle. The upward pressure of the cork pushing against the expanded glass neck and metal cap prevented the total expulsion of the cork from carbonation pressure. This device, although similar in principle to the wire ties of earlier days, allowed a fancier and potentially useful surface for advertising.

The Crown Closure

In 1889, William Painter of Baltimore, Maryland, began the development of a bottle closure which was to become a standard seal used from that day to the present. In a series of patent applications made in 1889, 1890 and 1891, and with a final one issued February 2, 1892, Painter developed the crown closure with which we are all familiar. This was not his first major patent, however, for Painter had developed earlier the New Bottle Seal (previously described) and was the holder of numerous other patents on stoppers of varying designs which were competitive in their day. His firm, known as the Bottle Seal Company, was reorganized in April, 1892, as the Crown Cork and Seal Company, and with a capital of $1 million he began a most successful venture in the soda water industry.

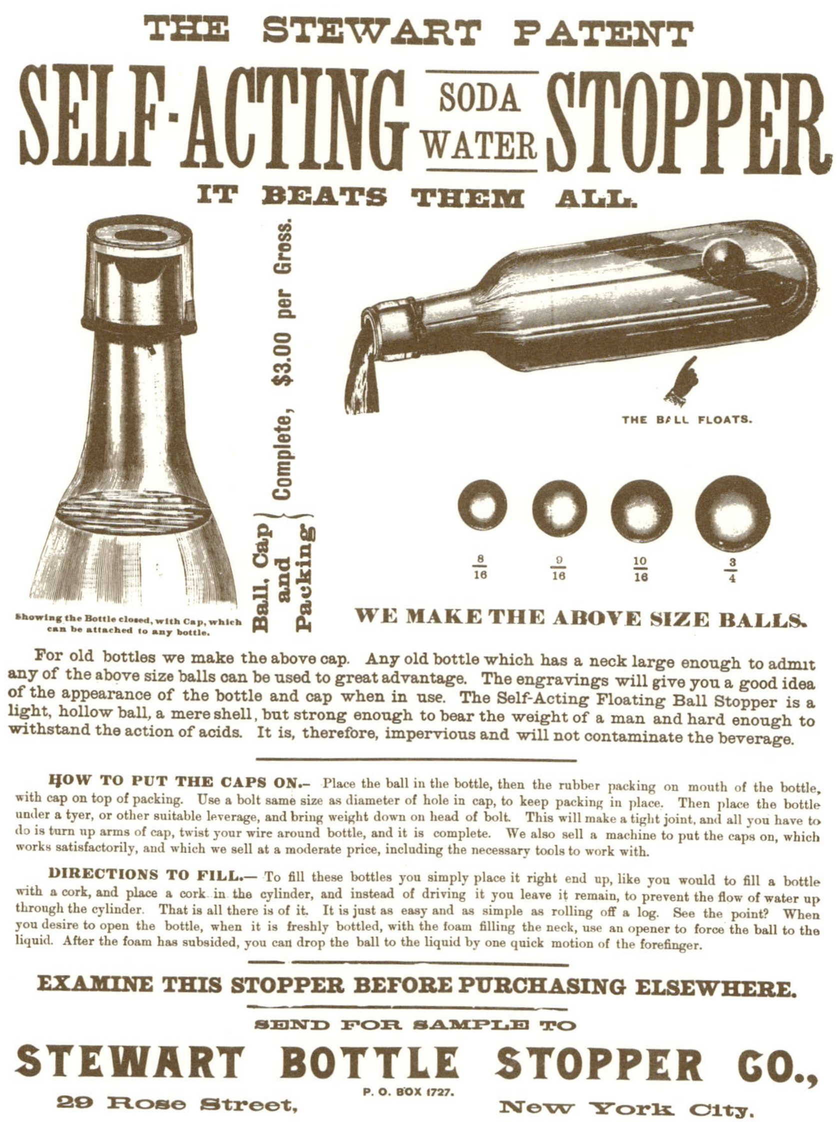

FIGURE 15. The internal screw-stoppered bottles (left, *N.B.G.* 1884) and floating metal ball stoppers (right, *N.B.G.* 1886) were widely advertised but seldom used by American bottlers.

HUTCHINSON'S

PATENT SPRING STOPPER.

Cheapest, Most Durable, Best and Simplest in Use.

OUR REFERENCES ARE 2000 BOTTLERS

Of the United States who use them. Our claims for them are Economy, Durability, Cleanliness, Ease with which they can be Used and Opened, and the General Satisfaction which they give to Customers.

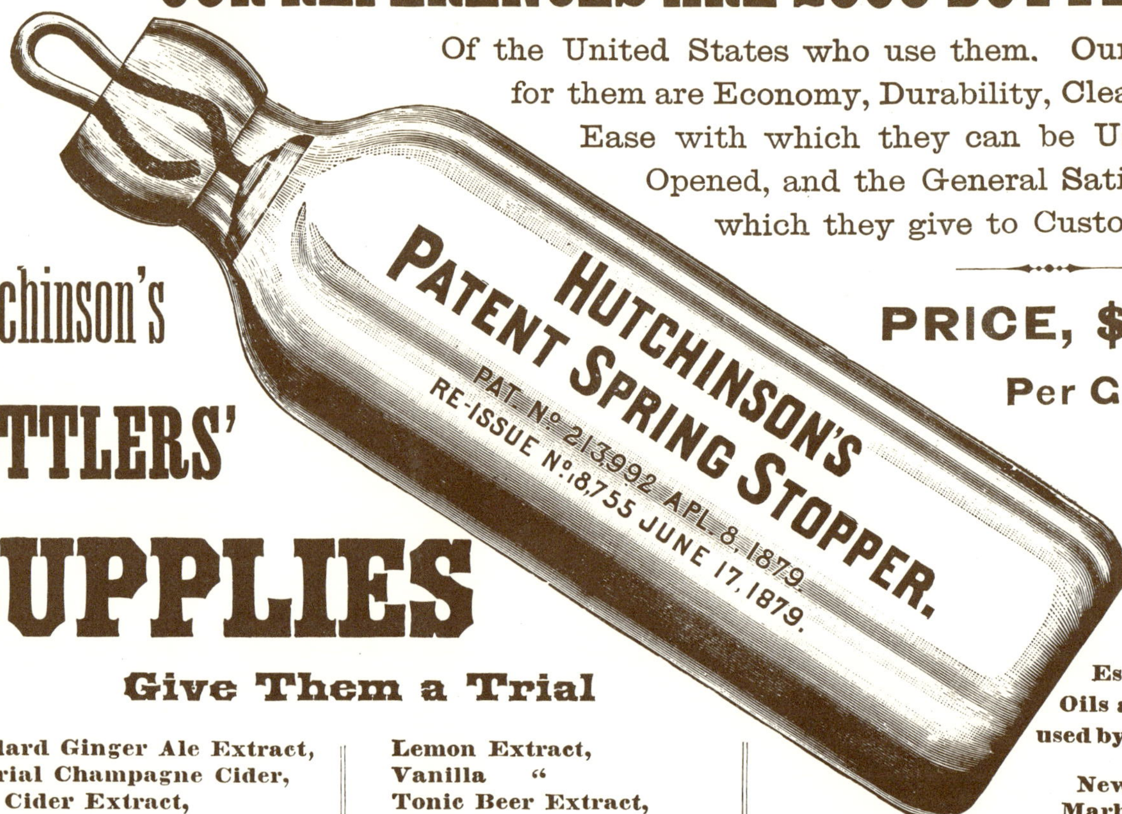

PRICE, $2.50

Per Gross.

Hutchinson's BOTTLERS' SUPPLIES

Give Them a Trial

Standard Ginger Ale Extract,
Imperial Champagne Cider,
Pear Cider Extract,
Apple Cider Extract,
Mead Extract,
Cream Soda Extract,
Club " "
Strawberry "
Raspberry "
Pineapple "
Sarsaparilla "
Soluble Lemon Extract,

Lemon Extract,
Vanilla "
Tonic Beer Extract,
Birch " "
Root " "
Little Daisy "
Spruce Beer "
California Pop Extract,
Strawberry Coloring,
Sugar Coloring,
Silver Foam,
Lime Juice.

Essential Oils and Acids used by the trade.

New York Marble Dust.

Ginger Ale Caps, Seltzer Caps, Tin Foil, Wire Cork Fasteners, etc.

Ginger Ale Loops and everything else.

Mineral Water Salts used by the trade.

SEND FOR PRICE LIST TO

W. H. HUTCHINSON & SON,

OFFICE AND FACTORY,

196 & 198 SOUTH DESPLAINES STREET, CHICAGO

And 60 PARK PLACE, NEW YORK.

W. G. SMITHERS, Manager.

FIGURE 16. The W. H. Hutchinson firm, best known for its stoppers, sold all the standard materials needed by the small bottler (*N.B.G.* 1886).

PARKHURST VS. HUTCHINSON!

PARKHURST.

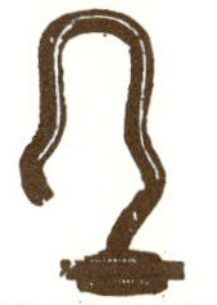

(Patented Dec. 11, 1883—No. 289,928.)
Was Invented by Amos F. Parkhurst,
PRIOR TO 1878.
It is manufactured as here shown, in accordance with claims of said patent.

TO BOTTLERS

AND THE

General Trade!

HUTCHINSON.

Patented April 8, 1879.
Re-issued June 17, 1879.

I am manufacturing, advertising and selling a Stopper called the Parkhurst, protected by a patent dated December 11, 1883, granted on the application of Amos F. Parkhurst, filed August 17, 1883. (See above cut marked Parkhurst). The Stopper manufactured, advertised and sold by W. H. Hutchinson & Son is an exact imitation and infringement of the Parkhurst patent, notwithstanding their assertion that they have a right to make same under a patent granted April 8, 1879, re-issued June 17, 1879. (See above cut marked Hutchinson). Any one comparing the cut of the Parkhurst Stopper shown above with the cut of the Hutchinson Stopper as shown above, will see that the Parkhurst patent is *NOT AN EXACT IMITATION* of the Hutchinson Patent.

It can readily be seen that the Patent Office, in December, 1883, did *NOT* grant a patent to Parkhurst for exactly the same thing a patent had been granted to Charles G. Hutchinson, in April, 1879. In Hutchinson's claim the wire is bent to approximate the figure 8. in form, whereas the Parkhurst claim reads in full as follows:

"The combination, with the neck of a bottle, of an internal stopper, provided with a laterally-yielding spring-bail formed of a continuous piece of wire, rigidly secured by one end to the stopper and located in and so arranged as to act in conjunction with the neck of the bottle to hold the stopper suspended in either an open or a closed position, substantially as described"

It will be observed that Parkhurst does not embody in his claim a bail approximating the figure 8 in form or a bail crossed.

Neither am I manufacturing such a stopper, and claim that I do not in any way infringe any of the Hutchinson patents, and to substantiate my position I quote from the Examiner of Patents, at Washington, as follows:

"Parkhurst has drawn no claim which interferes with claim of Hutchinson."

Hutchinson is, and has been, making, advertising and selling Stoppers of a device **that is not shown in any of the Hutchinson Patents,** and in view of said fact he sometime since brought suit in Chicago, Ill., to test the validity of my **PARKHURST PATENT, which shows, describes and claims just such a stopper as the Hutchinsons have been making, advertising and selling.**

Hutchinson is therefore infringing the Parkhurst Patent. He never brought suit under his patent for infringement, his assertion to the contrary notwithstanding, as his patent is for a different device, beside being of **later date** than the **Parkhurst Invention.**

Will add, that parties who use an infringing device are liable as well as the manufacturer.

Those purchasing the Parkhurst Stopper from me will be fully protected against parties claiming same an infringement on any patents.

See Advertisement of W. H. Hutchinson & Son, in former issues of The Bottlers' Gazette.

EDWARD H. EVERETT.

I am prepared to furnish PATENT BOTTLE STOPPERS at Lowest market price; also, STOPPER FINISHED MINERAL WATER and LAGER BEER BOTTLES.

EDWARD H. EVERETT,

Glass Manufacturer, NEWARK, OHIO.

FIGURE 17. Fights over patent infringements were common and Hutchinson defended his stopper against many competitors, such as the Parkhurst sold by E. H. Everett (*N.B.G.* 1886).

FIGURE 18. The Putnam firm was best known for its Lightning stopper, but it also distributed several other types of swing stoppers (*N.B.G.* 1901).

Many problems faced Painter as he developed the first of the crown stoppers, especially in holding the crown to the bottle and seating the crown with sufficient force to completely seal in the carbonation. By experimenting with crowns of varying sizes and cork liners of varying thicknesses, Painter was able to develop a tight and dependable seal for virtually all beverages.

Early advertising of the "crown cork system," as it was then known (Fig. 19), enumerated the merits of this new-style closure. Perhaps its major advantage, compared with all previous types of seals, was its security against being accidentally damaged or opened during normal handling and shipment. Only through direct effort could the bottle be opened, and such acts were not likely to occur in the normal handling procedures. Some of the early ads also indicated, with some naivety, the ease with which the bottle was opened. The promotion of such dubious techniques as the utilization of a scissors blade, table knife, or corkscrew probably led to many cut hands and fingers and considerable verbiage of an ungentlemanly nature. In order to ease the difficulty, however, the company from the very start offered 40 gross of bottle openers with each 1,000 gross of crowns ordered.

Despite the apparent desirability and advantages of the crown cork system, this closure was not immediately used by all firms. Competition continued from cork suppliers and users of the Hutchinson stoppers, the swing stoppers and the internal devices. Although the necessary modifications in bottle manufacture were relatively modest compared to the advantages of the crown closure, early glass bottle manufacturers were somewhat hesitant to retool until the bottling trade was ready to accept the new product. Also, the logistical task of adapting every bottler's plant to this new style was certainly a tremendous task. As late as 1905 probably less than 25 percent of the bottlers were using the crown; but within the next 10 to 20 years, bottle manufacturers made wholesale transformations of their finishing techniques to adapt to this seal of the future.

An additional advantage and selling point of the crown cap was the immediate availability of a new advertising space on the top of the crown. For a few cents per thousand, each local bottler was able to have individualized caps for his product, allowing ready identification of the contents and the identity of the bottler as well. Even the firm of W. H. Hutchinson & Son quickly added the crown cap to its array of bottling supplies and became one of the major distributors for this now totally acceptable closure.

One of the last minor changes or variations of the crown closure was the development of the "Priof" method

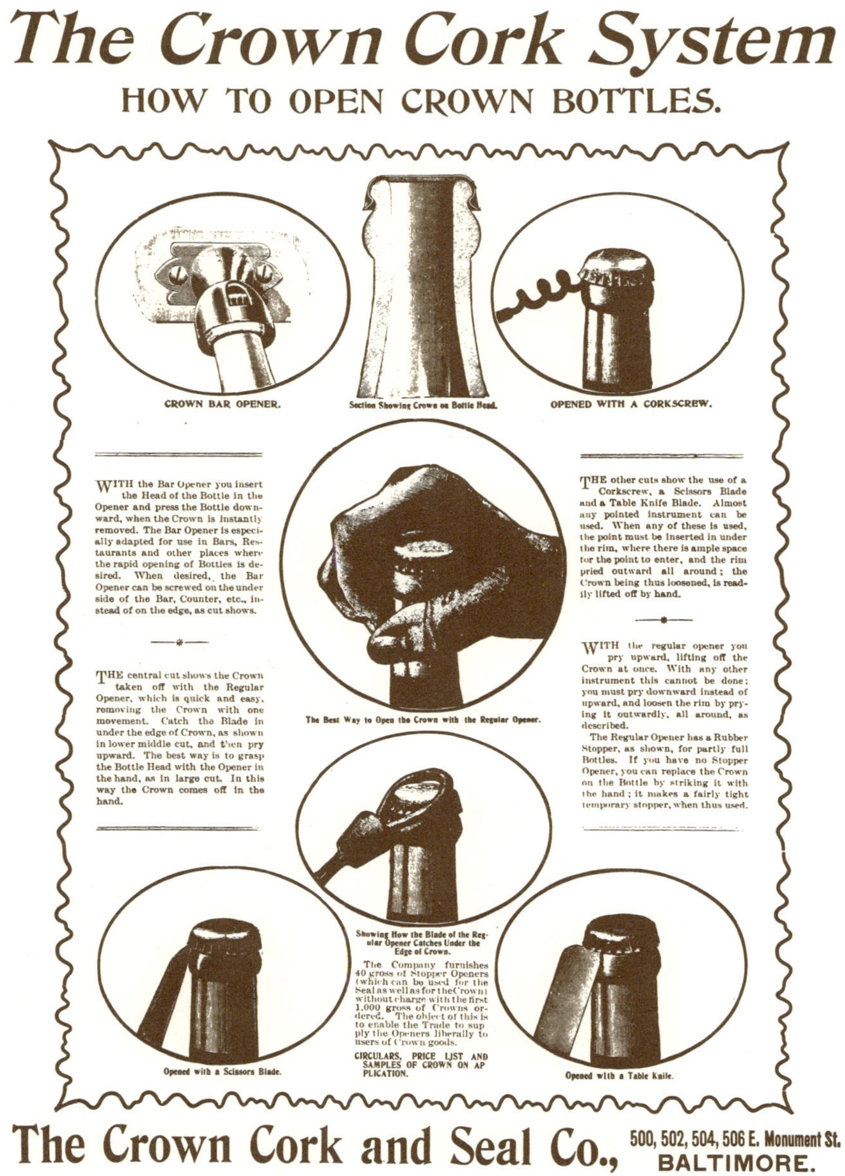

FIGURE 19. William Painter of Baltimore invented the "New Bottle Seal" (right, *N.B.G.* 1886) and the crown cork system (left, *N.B.G.* 1896).

FIGURE 20. The Bernardin Bottle Cap was a device used to help retain the cork in various carbonated beverages (*N.B.G.* 1886).

FIGURE 21. The "Priof" finish was a modification of the basic crown style which eased removal of the cap by providing a shoulder as a point of leverage.

of bottle finish by the Illinois Glass Company. The Priof crown bottle (Fig. 21), appearing in the early 1920s, had a projecting ledge below the lip of the crown which assisted the consumer in opening the bottle. By using the ledge as a point of leverage, one could easily pry off the cap by using a knife, key, coin or any flat piece of metal. One needed only to insert the lever between the ledge and the cap and twist. Although this type of bottle finish was used sporadically through the years, it never became popular enough to dominate the market. However, its major advertising feature was that it substantially reduced the number of broken and chipped bottles which resulted from the use of various unconventional openers.

Chapter IV
BOTTLES AND BOTTLE MANUFACTURERS

The rate of growth experienced by the soft drink industry in Illinois, and throughout most of the eastern United States for that matter, from 1885 to 1920 was tremendous. A combination of factors, including the development of new flavorings, more and varied advertising, refined methods of production and expanded distribution of the product, brought about such phenomenal growth. Without a readily available source of bottles, however, expansion of the industry—especially on a local level—would have been drastically curtailed. Glass manufacturing companies, quick to realize the potential market for soda water bottles, increased their production and advertised a variety of styles and sizes that would satisfy the needs of any bottler of the times (Figs. 22 and 23).

General acceptance and adoption of the Hutchinson stopper in the early 1880s (see Chapter III) was one of the basic steps that led the way to a more rapid and economical bottling procedure. Prior to this, soda pop and mineral water were contained in bottles sealed with corks or other forms of simple plugs. The applied neck and lip of these early bottles, often referred to as the "true blob," were designed to take the stopper and wire holder. However, should the wire break or become bent or the stopper become ineffective with age or wear, the entire device had to be replaced. Pre-Hutchinson soda or mineral water bottles are now relatively rare, mainly because of the time that has elapsed since their period of use and because, compared with the styles used after 1880, they were few in number.

An almost immediate change in the style of soda pop bottles took place with the development of the Hutchinson stopper. In order to accommodate the simple wire loop and rubber disk that formed the basis of this internal stopper, the mouth or lip of the bottle was applied as a thick, heavy, rounded top. The mouth and neck had to be capable of withstanding the tension exerted by the wire stopper when it was inserted in place after the filling operation and the pressure of the disk against the constricted neck after carbonation sealed it in place. This form of bottle, used almost exclusively for soda pop during the period from about 1885 to 1910, is universally referred to by collectors as the "Hutchinson blob."

Although most manufacturers advertised several sizes and styles of this type of soda pop bottle, the majority of bottles used were cylindrical in shape and nearly all the same size. Fluted bottoms appealed to some bottlers, but this form of paneled bottle made up only a small percent of the total. A Hutchinson blob style of even greater rarity has all sides, from shoulder to bottom, paneled; the Union Bottling Works of Mattoon and the Enterprise Bottling Works of Lincoln appear to have been the only two Illinois firms using this style. The majority of these bottles had a capacity of 6½ to 8 ounces, or roughly ½ pint. Today's automatic bottle-filling process is precise and the content of all bottles is nearly uniform; the method of hand-filling the Hutchinson soda bottle was not always so exact, however, and the contents in bottles from any given case of soda might vary considerably. The average Hutchinson blob measured 6½ inches in height with a base diameter of 2¼ inches. Some bottle manufacturers advertised pints and quarts (Fig. 22), but thus far few blobs of this latter capacity have been found that were unquestionably used for soda by Illinois firms. It was not until crown soda bottles came into general use that the quart bottle (Fig. 24) became popular.

The Illinois companies of E. H. Doolin of Jacksonville, C. A. Tolle of Litchfield, and the Lincoln Bottling Works of Lincoln are examples of the relatively few firms which purchased clear glass bottles. Probably 90 percent of all Hutchinson blob soda bottles used in Illinois were some shade of aqua, the intensity of greens and blues exhibiting considerable variation. Although the use of amber bottles came into favor for a short time with the early development of the crown soda bottle, only one Illinois bottler, the Star Bottling Works of Westville, is known to have used an amber pre-Hutchinson blob for soda. Cobalt blue soda pop bottles are also extremely rare; two of the very few Illinois firms known to have used them were John A. Lomax and W. H. Hutchinson of Chicago.

There were numerous glass manufacturing companies flourishing in the eastern United States during the last quarter of the 19th century, and in an effort to satisfy the demand for ever-increasing numbers of bottles from the growing soda pop industry, many expanded their plants and diversified the products sold. The so-called "plate-mold" came into greater use, and as a result a bottler could order embossed bottles with wording of his choice. The glass company simply inserted a replaceable metal plate, stamped in relief with the ordered lettering, into a standard mold before the molten glass was blown into it. The outline of these plates is usually visible on the bottle as a circle or half-circle bordering the lettering. Some molds, however, were made with the requested wording stamped directly into the sides and/or bottom of the mold itself in which case a plate was not used. This method of embossing bottles was continued until the coming of the decoratively embossed proprietary bottle in the early 1920s. The most common wording arrangement was circular in outline, although a few companies, such as J. J. Flynn of Quincy, ordered their bottles — the early varieties, at least — embossed vertically (See Fig. 81).

Embossing was not always restricted to the bottler's name or company, city and state. Many bottlers, especially those in the larger cities who registered their company and/or a particular trademark, often included this information as well as the actual trademark itself. These

FIGURE 22. This ad for the Whitney Glass Works (*N.B.G.* 1887) shows the variety of available blob-top bottles.

FIGURE 23. The Illinois Glass Company was still manufacturing Hutchinson blobs as late as 1910 as shown in this *N.B.G.* ad.

FIGURE 24. Quart soda and mineral water bottles used by Illinois firms from about 1900 to 1920. The three center bottles in the middle row are half-gallons.

trademarks varied from such things as ships, animals and anchors to elaborately superimposed letters, usually initials of the bottler's name or firm (Fig. 25).

Embossing served two basic purposes: first, it identified the bottle as belonging to a particular individual or company and, second, it provided a means of advertising. Many bottlers had their name or initial embossed on the bottom of the bottle (Fig. 26), and since the Hutchinson bottles were stored and shipped "bottom-side-up" in crates, they could be identified immediately in that position. One firm, M. A. Fischer of Springfield, also had the letter F embossed on the shoulder of its Hutchinson-style bottles; whether or not this was done for the purpose of identifying the company's bottles is a matter of speculation.

The matter of theft and loss of bottles was a major concern to all bottlers during this period. It was not until the 1901 annual convention of the American Bottlers' Protective Association, founded in 1889 for the principal purpose of protecting the bottlers' property rights in their bottles, that serious consideration was given to the matter of establishing a generally acceptable system of charging a deposit on bottles. Up to that time bottlers depended entirely on the customer's voluntary willingness to return the empty bottles. Apparently some firms began early to charge a bottle deposit, but the practice did not become widespread until the late 1920s. In an effort to discourage customers from simply throwing away the empty bottles and competitive firms from stealing or at least using another's bottles, embossed statements warning against such practices came into general use. Statements such as the following are often found on Hutchinson blobs and beers and on early crown sodas of the period 1905 to 1920:

THIS BOTTLE IS NEVER SOLD

THIS BOTTLE IS LOANED, BUT NEVER SOLD, RETURN IT

THIS BOTTLE NOT TO BE SOLD

THIS BOTTLE REGISTERED AND NEVER SOLD

LOANED — NEVER SOLD

THIS BOTTLE NEVER SOLD OR ABANDONED

Perhaps the statement embossed on an applied crown soda bottle of the A.M. & B. Co., (Waco, Tex., and St. Louis, Mo.) best reflects the feeling of many bottlers in business at the turn of the century: WE PAY FOR EVIDENCE CONVICTING THIEVES FOR REFILLING OUR BOTTLES. This was no idle threat, and there was more than one bottler who brought a competitor to trial for just such an offense. In the October 5, 1901, issue of the *National Bottlers' Gazette*, the following statement appeared:

> The Springfield [Illinois] Mineral Springs Company, on September 18 seized a ginger ale case under a writ of replevin which was being shipped to A. Davis at Peoria, by Sam Davis, his son, of this city. Justice Brinkerhoff issued the warrant. Sam Davis was formerly an employee of the company. The Mineral Springs Company recently secured an injunction restraining W. J. Grant from taking and filling its bottles, and the proprietors say they intend breaking up the practice that has so long prevailed here of one bottler using the property of other bottlers whenever he can pick such up.

Development of the crown soda pop bottle and the cork-lined metal cap during the 1890s brought about an almost complete replacement of the Hutchinson bottle by 1915. Although glass manufacturers were still advertising the Hutchinson blob as late as 1910 (see Fig. 23), and it was used by a few bottlers for awhile after that, the crown soda was soon to be universally accepted and used. Until the advent of the automatic bottling machine in 1903 and its installation in glass manufacturing plants, the upper neck and mouth of the crown bottle was applied by hand in a process not unlike that used for its forerunners.

In actual use the automatic bottling machine was not operational for the production of pop bottles for some time, and it was not until about 1915 that most bottles were so manufactured. The Illinois Glass Company (Alton, Ill.) states in its 50th anniversary report that the first successful bottle-blowing machine was made in 1906 and that they installed their first in 1910. Among the difficulties involved in installation of such an apparatus was the resulting unemployment of many hundreds of glassblowers and workmen. The American Bottling Company plant (Streator, Ill.) experimented with the semi-automatic machines in 1900 and added 12 of the Owens automatic machines in 1906. However, they continued for some years with their hand-blown operations. In 1908 several additional machines were added to their lower works. But these still did not displace the last of the hand-blown bottles.

Embossing was used by most firms on the crown bottles as it had been on the earlier blob sodas, and there was a tendency now to include the amount (for example, "7½ Fluid Ounces") of the contents. Most of these bottles were similar in size and outline; one notable exception was the "bowling pin" bottle employed by the Gravel Springs Co. (Jacksonville, Ill.) (Fig. 27). It was not until the desire for more attractive and distinctive bottles was made known that the rather plain bottles were replaced by a wide variety of specially designed and embossed crown sodas. The early 1920s saw the beginning of the so-called proprietary bottle (Fig. 28).

These proprietary bottles, produced on automatic bottling machines and involving complex molds, were more elaborately designed than their crown predecessors. They quickly found favor with local bottlers, many of whom took advantage of the opportunity to order bottles with their own name and trademark — such things as an Indian head, wolf head or a prairie chicken — plus an overall detailed pattern. These designs in the glass consisted of various combinations of lines, bars, panels, diamonds, stars and the like (Figs. 29 and 30). Several of the larger companies developed and patented their own particular styles of bottles, and each style soon became identified with a particular company (i.e., a proprietary style). Some of these bottles bear the patent date awarded for a special design or style: Orange Crush, July 20, 1920; Smile, July 11, 1922; Chero Cola, June 3, 1924. The basic bottle design still in use by the Coca-

FIGURE 25. Trademark emblems used by early Illinois bottlers included initials, animals and other objects.

FIGURE 26. Initials, names and other symbols were embossed on the bottoms of Hutchinson and early crown soda bottles to establish and facilitate identification.

FIGURE 27. The "bowling pin" style of bottle used by the Gravel Springs Company of Jacksonville, Illinois, ca. 1905-1915.

Cola Company was first patented November 16, 1915, the forerunner in the development of proprietary bottles. A few years later slight design modifications received a patent on December 25, 1923; bottles bearing this date have been termed "Christmas Cokes."

The paper label had been in general use for a long time in connection with the early embossed bottles. This was an economical means of identifying the contents and it served as a supplement to the embossing by providing additional information (see Chapter VI). However, paper labels had to be replaced after each washing, an added expense in the refilling and capping process. The problem was solved in the mid-1930s with the development of the applied color label (Fig. 31). Under normal conditions, the applied color label proved to be exceptionally durable and lasted the life of the bottle. Many of the early, more elaborately designed proprietary bottles were phased out of use and the general style was simplified to accommodate the painted label. The evolution of the soda pop bottle reached its climax, for the most part, in those forms in general use today — a combination of limited embossed lettering, some glass design, and large painted labels.

Several glass-manufacturing companies were operating in Illinois during the latter part of the 19th century, and some produced beverage bottles in addition to their standard products of window glass, tableware, medicine and food bottles, fruit jars, and miscellaneous containers. However, it appears that the majority of Hutchinson and early crown soda bottles used by Illinois bottlers were purchased from companies outside the State, especially Indiana and Pennsylvania, despite higher shipping charges. Probably some of these eastern firms, such as Lorenz and Wightman of Pittsburgh, were first in producing quantities of soda bottles for wide distribution. The soft drink industry expanded rapidly during the last two decades of the 1800s and, as the Hutchinson blob came into general use, several long-established glass factories advertised a diversified line of soda bottles. Various styles could be purchased in quantity for less cost from glass companies in Pittsburgh, for example, than from some local firms.

Glass manufacturers often embossed their complete company name or initials on the bottles, usually on the "heel" or side just above the bottom. Not all companies followed this practice and there are a great number of extant bottles without a maker's name or initial. These initials prove invaluable in identifying the company that manufactured a particular bottle and, with the knowledge of a glass firm's period of operation, it is often possible to estimate the approximate time a bottler was in business. Since many of the early local bottlers have long ago gone out of business and little is known of the extent or time of their operation, the quantities and types of bottles used by them (and marked with the manufacturer's initials) provide some information of the enterprise. In addition, nearly all early bottles were embossed with the name of the bottler or the company, the city, and usually the state and, as a result, have become especially interesting as collectors' items.

FIGURE 28. Glass companies such as Reed offered the small bottler a ready-made proprietary style as well as custom manufacture of private mold bottles (*N.B.G.* 1925).

FIGURE 29. Characteristic proprietary bottles used by Illinois firms during the 1920s and 1930s.

FIGURE 30. Characteristic proprietary bottles used by Illinois firms during the 1920s and 1930s.

FIGURE 31. Painted-label bottles, a new method developed in the mid-1930s.

Published records of many of these early glass-manufacturing companies appear in the form of advertisements, news events, business transactions and various related items in the *National Bottlers' Gazette* and other trade journals. Owens-Illinois (Illinois Glass Company) published a special anniversary volume dealing with 50 years (1873-1923) of the company's history and operations. The most comprehensive reference available on this subject, *Bottle Makers and Their Marks* by Julian H. Toulouse (Thomas Nelson, Inc., Camden, N.J., 1971), will prove invaluable to those especially interested in this aspect of bottle manufacturing or collecting. The following brief summarization is based on information extracted from the above-mentioned and other related sources. The companies listed here were selected as a result of their names and/or initials being found — some rarely, many commonly — on the blob and early crown soda bottles utilized by Illinois bottlers (see also Chapter X).

A B Co

AMERICAN BOTTLE CO., Chicago, Ill., 1905-1916; Toledo, Ohio, 1916-1929.

This company was formed originally as a merger of the Adolphus Busch Glass Manufacturing Co. (Belleville, Ill. and St. Louis, Mo.), the Streator Bottle & Glass Co. (Streator, Ill.), and the Ohio Bottle Co. (Newark, Ohio) which was made up of several subsidiary plants in Ohio (Toulouse, 1971:30). It was a major producer of beverage bottles, including beers, but prohibition caused a severe cutback in production and the eventual closing of the Streator plant. Semiautomatic and automatic Owens machines gradually replaced the handmade bottle operation, the latter method being completely phased out by 1917. Hutchinson blobs and applied crown soda bottles (A B Co embossed on the heel) are known to have been purchased from the American Bottle Co. by several Illinois bottlers.

A & D H C

A. & D. H. CHAMBERS, Pittsburgh, Pa., 1843-1886 or later.

Toulouse (1971:38) indicates that, in its latter years of operation, this firm developed a large shipping business and marked its bottles with the initials A & D H C. Although no mention is made of soda bottle manufacturing, numerous Illinois firms purchased bottles from this glass company.

A G W, A G W L

AMERICAN GLASS WORKS, Pittsburgh, Pa., about 1865-1880.
AMERICAN GLASS WORKS LIMITED, Pittsburgh, Pa., about 1880-1905.

Hutchinson blob soda bottles embossed A G W or A G W L have been noted as being used by several Illinois bottling firms. The company advertised a large variety of styles and colors of beverage bottles in the *National Bottlers' Gazette* during the late 1880s.

C G Co

COSHOCTON GLASS CO., Coshocton, Ohio, 1907-1923.

The Coshocton Glass Co. was one of several producers of beverage bottles that, because of its specialization in beers, was forced to close as a result of prohibition. The company supplied bottles to several major breweries in both this country and abroad. It also produced an extensive line of soft drink bottles; however, only one Illinois firm, the Decatur Bottling Works, has been noted thus far as having used containers (Hutchinson blob) manufactured by this company.

C & Co

CUNNINGHAMS & CO., Pittsburgh, Pa. 1879-1907.

D O C

D. O. CUNNINGHAM GLASS CO., Pittsburgh, Pa., 1882-1937.

In 1882 Dominic Cunningham, the son of Wilson Cunningham who was one of the founders of Cunninghams & Co., formed his own company, the D. O. Cunningham Glass Co. (Toulouse, 1971: 120). The company soon began production of beverage bottles and in 1885 was advertising mineral, ginger ale and beer bottles. Many Illinois firms utilized bottles made by this company. Of the two manufacturers, C & Co. and D O C, bottles from the latter appear to be the most numerous.

E H E

EDWARD H. EVERETT GLASS CO., Newark, Ohio, 1883-1904.

This is another glass manufacturing company that grew rapidly as it strove to compete with other firms in supplying bottles for the expanding beverage industry during the late 1800s. Toulouse (1971:185-186) mentions the difficulty encountered by Everett with the W. H. Hutchinson Co. over the Parkhurst stopper (patented 12/11/1883). Everett preferred to use it, but Hutchinson claimed this infringed on his own patented stopper (patented 4/8/1879). Everett lost both court battles over its continued use, but he did gain considerable publicity through ads in the *National Bottlers' Gazette*. By 1900 the company was operating six furnaces for beer, soda and liquor bottles; but only a limited number of Illinois bottles have been found with this company's mark. In 1904 Everett formed the Ohio Bottle Co. which later was bought by the American Bottle Co.

F A & Co

FAHNSTOCK & ALBREE, Pittsburgh, Pa., 1860-1862.

This company, which leased the factory of Lorenz and Wightman for the two-year period, is best known for its extensive production of historic flasks. The "Union and Clasped Hands, reverse Eagle" and the "Union and Clasped Hands, reverse Cannon" were two of the most famous designs manufactured during this short period. Although some bottles were produced for soft drink bottlers and some apparently for druggists, flasks appear to have been the principal line (see also L & W, Lorenz & Wightman).

I G, I G Co, I

ILLINOIS GLASS CO., Alton, Ill., 1873-1929.

Numerous Illinois bottlers purchased from this famous glass manufacturing company, partly because of its location and partly because of the diversified styles and sizes of bottles it offered. The business expanded rapidly, employing over 60 men in 1876. By 1909 the operation had grown in size from one furnace to 11, and in that year the whole plant was rebuilt to accommodate the Owens machines (Toulouse, 1971:264). By 1915 thirty-one automatic bottling machines were in use and 2,400 workers were employed.

Bottles embossed with I G were manufactured before 1890; those marked I G Co from about 1880 to 1900; those with the I G Co enclosed in a diamond from about 1900 to 1916; and those with the letter I enclosed in a diamond from 1916 to 1929. Additional information on the development, expansion and mergers of the Illinois Glass Co. may be obtained by reading Toulouse (1971); "Fifty Years of Achievement in Building up a Service of Better Bottles," published by the Illinois Glass Company (1923); and other articles listed in "Suggested References" at the end of the text.

L & W

LORENZ & WIGHTMAN, Pittsburgh, Pa., 1851-1860, 1862-1871.

A few Illinois true blob soda bottles embossed with the large letters L & W are known. Fahnstock & Albree leased the Lorenz & Wightman glass works from 1860 to 1862 (Toulouse, 1971:338) and produced bottles embossed with their company initials, F & A and F A & Co. Because of the limited number of bottlers in operation in Illinois prior to 1870, bottles known to have been purchased from these glass companies are rare.

L G Co

LOUISVILLE GLASS WORKS (or CO.), Louisville, Ky., 1855-1873.
LOUISVILLE KENTUCKY GLASS WORKS, Louisville, Ky., 1873-1886 (or later).

Only one Illinois soda bottle, a true blob embossed with the name Edward Gundlach (Columbia), has been recorded thus far bearing the initials L G Co. The company is known to have produced various historical flasks, but the extent of the beverage bottle production is unknown.

M B & G Co

MASSILLON BOTTLE & GLASS CO., Massillon, Ohio, 1900-1904.

Only one Illinois firm, the Twin City Bottling Works of Urbana, was noted during this study as having utilized bottles (applied crown) from this short-lived glass bottle manufacturer. Massillon Bottle & Glass merged with the Edward H. Everett Glass Co. (Newark, Ohio), Reed & Co. (Massillon, Ohio), and the Wooster Glass Co. (Wooster, Ohio) to form the Ohio Bottle Co. In 1905 this newly formed company merged with the Streator Bottle & Glass Co. and the Adolphus Busch Glass Manufacturing Co., forming the American Bottle Co. (Toulouse, 1971:348).

McC, Mc & Co, Wm. McC & Co

WM. McCULLY, Pittsburgh, Pa., 1832-1841.
WM. McCULLY & CO., Pittsburgh, Pa., 1841-ca.1886.

The Wm. McCully glass manufacturing company, renamed the "Phoenix Glass Co." by the owner, is another Pittsburgh firm remembered for its production of historic flasks and, to a lesser extent, fruit jars (the "Dictator" and the "Magic"). Toulouse (1971:352) states that "Phoenix grew to be one of the largest glass plants of its day, shipping to the Midwest and to foreign countries"; but no mention is made of soda bottle production. Only one Illinois true blob soda (C. Dorries & Brother, Breeze) embossed Wm McC & Co was noted during this study.

M G Co

MODES GLASS CO., Cicero, Ind., 1895-1904.

W. F. Modes worked for the De Steiger Glass Co., D S G Co (La Salle, Ill.) and the Streator Bottle & Glass Co., S B & G Co (Streator, Ill.) before starting his own company in Cicero, Indiana. The factory burned in 1904 (Toulouse, 1971:360) and Modes rebuilt it under the name Indiana Bottle & Glass Co. He also started the Olney Bottle Co. (Olney, Ill.) in 1910, but it lasted only four years. Bottles produced at his factory in Cicero bear the initials M G Co. A Hutchinson blob from Litchfield (William Gerhardt) is the only Illinois soda bottle recorded thus far as having been made at the Modes Glass Co.

N

OBEAR-NESTER GLASS CO., East St. Louis, Ill., since 1894.

Apparently the Obear-Nester Glass Co. produced some beverage bottles during the earlier years of its operation although no Illinois Hutchinson or applied crown soda bottles bearing this firm's trademark have been encountered thus far. After 1920, however, the volume of production was increased considerably and included a variety of containers such as liquors, beers, prescriptions, vials and proprietary sodas (Toulouse, 1921:375). At least five Illinois bottlers purchased some of their bottles from this glass company; all are the clear, proprietary type produced on automatic bottling machines and bear the company's N monogram in a square, a style adopted in 1915. One of these bottles, from the Mester Bottling Co. of Springfield (Ill.), has the patent date of June 3, 1924, embossed on the bottom.

N B B G Co

NORTH BALTIMORE BOTTLE GLASS CO., North Baltimore, Ohio, 1885-1895; Albany, Ind., 1895-1908; Terre Haute, Ind., 1900-1930.

The North Baltimore Bottle Glass Co. was one of the major suppliers of soda and mineral water bottles during the years of greatest growth of the soft drink industry. The company specialized in beverage bottles that were "strictly high grade" (Fig. 32), and, although the 1925 advertisement illustrates strictly crown sodas, the company also produced great quantities of the Lightning and Hutchinson stopper bottles. A great many Hutchinson blobs embossed N B B G Co have been noted, and it is apparent that numerous Illinois bottlers patronized this company. It is of interest to note that Toulouse (1971:380) attributes the company's eventual closing to prohibition and to the fact that it could not compete due to its failure to install automatic bottling machines.

FIGURE 32. An advertisement for bottles from the North Baltimore Bottle Glass Company (NBBGCo) of Terre Haute, Indiana. This company manufactured glassware for many midwestern bottling firms (*N.B.G.* 1925).

R G Co

RENTON GLASS CO., Renton, Wash., 1907-1911.

Toulouse (1971:440-41) states that the Renton Glass Co. apparently did not identify its bottles during the first years of operation—that it was only near the closing of the plant that bottles were embossed R G Co. It is interesting to note that two Illinois bottlers, H. L. Hester (New Burnside) and Altamont Bottling Works (Altamont), purchased their bottle supplies from this firm so distantly located when numerous other glass companies were much closer.

ROOT

ROOT GLASS CO., Terre Haute, Ind., 1901-1932.

The first glass plant built by Chapman J. Root was for the production of beverage bottles while the second (1902-1909) was started for fruit jars. Toulouse (1971:445) states that the beverage bottles were handmade until 1912 when the Root plant was converted to semiautomatic machines developed by their own personnel. During this early period Root produced the Hutchinson soda bottle. Although blobs used by several Illinois bottlers have been noted with the embossed ROOT on the heel, crown sodas produced by this glass company are far more numerous.

Alex Samuelson, a foreman in the Root plant, developed a bottle design that was accepted by the Coca-Cola Company and which was patented November 16, 1915. After this event, most of Root's beverage bottle production involved the manufacture of Coca-Cola bottles. The Owens Illinois Glass Company purchased the Root Glass Company in 1932 but continued to use the Root name for the next two years.

S B & G Co

STREATOR BOTTLE & GLASS CO., Streator, Ill., 1881-1905.

The Streator Bottle & Glass Co. was one of the most successful Illinois glass manufacturers. Over 300 workers were employed by 1885, and before long the factory was operating 24 hours a day. The company began experimenting with imported semiautomatic bottling machines as early as 1900, and it remained a forerunner in the production of beer and other beverage bottles until merging with the American Bottle Co. in 1905. In spite of the company's production capabilities and its location in the state, very few soda bottles used by Illinois bottlers have been encountered that bear the S B & G Co embossing.

W F & S MILW

WILLIAM FRANZEN & SON, Milwaukee, Wis., 1900-ca.1923.

William Franzen was serving as president of the Northern Glass Works (formerly Northern Glass Company, Milwaukee) in 1900 when he took ownership and began using his own mark, W F & S MILW. According to Toulouse (1971:537), after Franzen became owner the quality of bottles improved and the Milwaukee brewers again became the factory's most important customers. High production costs and prohibition brought about its downfall.

Soda bottles were also manufactured, but apparently few were used by Illinois bottling firms. Only two bottles embossed W F & S MILW have been noted, a Hutchinson style Bristol Mineral Spring (Chicago) and a crown soda used by Clem Bussan of Galena, the latter apparently produced on a semiautomatic bottling machine.

WIS. G. CO. MILW.

WISCONSIN GLASS CO., Milwaukee, Wis., 1882-1886.

Although the period of operation of the Wisconsin Glass Co. was of short duration, a variety of glass products were manufactured including fruit and food jars, ink and beverage bottles, carboys, and window glass (Fig. 33). The company advertised "Private Moulds a specialty" and described its soft drink bottles as "extra strong" and able to "stand high pressure." Illinois soda bottles embossed WIS. G. CO. MILW. are rare, apparently because of their reportedly high cost and the short period of manufacture.

FIGURE 33. Advertisement for the Wisconsin Glass Company of Milwaukee, Wisconsin (*N.B.G.* 1884).

FIGURE 34. Examples of early blob-top soda bottles used by F. W. Ehrhardt of Beardstown, Illinois. The top row (a) shows three views of one bottle. The bottom row shows three additional types of bottles used by this firm in the late 19th century.

Prior to 1880 there were few sources from which to purchase soda bottles and supplies. Round-bottomed and pointed-bottomed (torpedo) bottles were favored by some bottlers, but these styles (Fig. 9) apparently were never used extensively in the Midwest. Occasionally imported soda bottles were purchased. One of the more interesting types involved the use of an internal glass or metal ball stopper (see Chapter III). These were undoubtedly expensive for that time, and thus far none have been found that could be associated with any Illinois bottling enterprise. One Illinois bottler, F. W. Ehrhardt of Beardstown, used several interesting styles of bottles during his operation, some made by U.S. glass companies and some of foreign make. The most unusual was a tall (9¼-inch), paneled bottle embossed vertically with the words, MANUFACTURED BY/F.W.EHRHARDT/BEARDSTOWN ILL/SUPERIOR/BELFAST/GINGER ALE, with each word or line on a separate panel (Fig. 34, A). A second type of imported bottle used by Ehrhardt was round, had the letter E embossed in a star on the bottom and the two lines BELFAST & DUBLIN/AROMATIC GINGER ALE embossed near the base (Fig. 34, B). How many of these imported bottles were used or for how long is unknown. A true blob used by Ehrhardt is also rather unusual for that period in that it is entirely panelled (10 panels) with the name F. W. EHRHARDT embossed vertically on one panel (Fig. 34, C). The later, more conventional Hutchinson blob is simply embossed F. W. EHRHARDT/BEARDSTOWN ILLS. (Fig. 34, D); both blobs had been made by the Illinois Glass Company of Alton and were embossed I G Co. (ca. 1880-1900).

MAJOR COMPANIES AND THEIR BOTTLES

The Coca-Cola Company

Probably no other American soft drink company has been as successful, remained in business for as long a period with the same trademark, or attracted as many individuals to collect its paraphernalia as has the Coca-Cola Company. Coca-Cola syrup was created and first sold in Atlanta, Georgia, in 1886 by its originator, John S. Pemberton; but it was not until 1894 that it was first bottled in Vicksburgh, Mississippi, by Joseph A. Biedenharn. Hutchinson blobs embossed with Biedenharn's name are well known, but these bottles do not bear the name "Coca-Cola." Hutchinson blobs embossed with the name "Coca-Cola" are extremely rare, but this is quite understandable since it was not the usual practice to state (emboss) the contents on these early soda bottles.

Coca-Cola bottling plants were established in many cities during the first decade of the 20th century. The crown soda bottle had been developed and was rapidly replacing the earlier Hutchinson blob at this same period. From about 1900 to 1915, when the present Coke bottle style was first adopted and patented, a variety of embossed Coca-Cola lettering styles appeared on the early crown soda bottles used by the company (Fig. 35). The first of these had applied crowns, but by about 1915 nearly all of the crown sodas used by Coca-Cola were full-mold bottles. For an unknown but apparently short

FIGURE 35. Coca-Cola bottles of the 1900-to-1920 period. The bottle on the right, patented in 1915, is the basic style in use today. The center bottle did not contain Coca-Cola but was used by a Coca-Cola distributor.

period of time, amber bottles were used prior to the adoption of the standard green Coke bottle; amber crown sodas were used also by the Charles E. Hires and Pepsi-Cola bottling companies as well as by Coca-Cola. Bottles possessing the words "Coca-Cola" embossed in script normally contained "Coke" while other bottles with block lettering such as THE COCA-COLA BOTTLING COMPANY might contain any one of several other flavored drinks bottled at a particular plant.

The Pepsi-Cola Company

Caleb D. Bradham, a pharmacist in New Bern, North Carolina, developed a soft drink formula in 1896 for use at his counters which he called "Brad's Drink." In 1901 he changed the name to "Pepsi-Cola" and, because the drink was so well received locally, Bradham formed the Pepsi-Cola Company two years later. Some of the earliest crown sodas used by the company were embossed with the word "Registered," and the name "Pepsi-Cola" in script on the sides or near the heel. An early proprietary bottle from Henderson, North Carolina, (manufactured by the Root Glass Company of Terre Haute, Ind.) has the name "PEPSI:cola" embossed in large block letters on the shoulder, plus the city and state, "Registered" and "contents 6½ fluid ozs" in smaller letters. Near the bottom is the embossed statement, "This bottle not to be sold under penalty of the law." Much later proprietary bottles of the 1930s and 1940s had the name "PEPSI-COLA" embossed vertically in panels around the shoulder with a paper label on the bottom half. Today the "swirl" bottles bear a simple painted label with the word "Pepsi" just above the shoulder.

The Charles E. Hires Company

Charles E. Hires began his career as an apothecary's apprentice in Millville, New Jersey, in 1869. Soon after, he moved to Philadelphia where he established his own drugstore business. It was simply a matter of chance that he encountered a homemade "herb tea" which had been concocted from native plant roots and berries by a farm woman in New Jersey. Hires obtained the recipe and, by a great deal of experimentation with and careful blending of the ingredients, developed an extract he considered an accurate facsimile of the original herb tea drink which he thought at the time had such commercial potential. His newly developed "Hires Root Beer Extract" was exhibited at the Philadelphia Centennial in 1876 with samples of the finished product served free to anyone interested in trying this new "temperance" drink.

Promotion of the extract was not an easy matter in those beginning years; but by the end of the first year after its formal introduction, an estimated 11,500 glasses of Hires had been sold to the public. Hires featured his new product as a temperance drink and it was occasionally promoted as a tonic to "purify the blood and make rosy cheeks." Nevertheless, the drink met bitter opposition during the mid-1890s from the Woman's Christian Temperance Union which contended that "the fermentation of a sweet liquid in ordinary temperatures always produces alcohol." Through honest representation of his quality product, however, Hires Root Beer Extract — and later the bottled drink — became the most popular of all the root beers.

Advertising played an important part, as it did in the promotion of all soft drinks produced by those com-

FIGURE 36. Three early applied-crown Hires bottles. The one on the left is embossed "Hires" on the bottom.

panies which became and stayed successful, in the general acceptance of Hires Root Beer (see Chapter VIII, Fig. 58). It was not until about the mid-1890s, however, that the Charles E. Hires Company began to advertise for bottlers to undertake the actual bottling and distributing of the root beer drink. By the end of the first decade of the 20th century, numerous private companies in the East were bottling and distributing Hires Root Beer, and it was during this period that the Hires Company began its own bottling operations as well. The first bottles directly associated with the company were tall (9½-inch), applied-crown sodas with simply the name HIRES on the bottom or the phrase HIRES CARBONATED BEVERAGES embossed on the side (Fig. 36). Some bottles of this latter type bear the letters W. G. CO. on the bottom (possibly the Whitney Glass Works of Glassboro, N.J. 1887-1918). With the coming of the proprietaries in the early to mid-1920s, Hires patented a basically plain style of proprietary that has changed little over the years — except for the addition of the painted label. Today Hires is the largest manufacturer of root beer in the world, and the company accounts for approximately 25 percent of all sales in this country.

CHAPTER V
FLAVORINGS AND EXTRACTS

The relatively rapid transition of soda water from a simulation of a mineral water for medicinal purposes to a refreshment drink was enhanced by the early use of flavorings. Riley (1958) suggests that it was perhaps the degree to which a variety of flavorings was used in the American soda waters that distinguished them from their European soft drink counterparts. A substantially large proportion of European carbonated waters were sold for medicinal purposes for many years, long after the popularization of flavored beverages in America. Among the earliest flavors were spruce beer, root beer, sarsaparilla and birch beer, all of which were made from an infusion of barks, sugar, water and yeast. A light fermented ginger beer, made of sugar, citrus juices, honey, ginger and water, was popular in the early years of the 1800s. "Mead," largely a concoction of honey and yeast, lightly fermented, was apparently popular in the 1820s and 1830s when a number of patents were issued for its manufacture. One modification of this drink was a carbonated "Sarsaparilla Mead" for which a patent was issued in 1824 to a John L. Schieffelin of New York City.

Mead, spruce beer and tonic root beer were bottled by George S. Twitchel in Philadelphia about 1850, and spruce beer was the first product manufactured by William Hutchinson when he opened his plant in Chicago in 1848. The alcoholic content of these small beers was about two percent when made, rising to three or four percent when aged.

Citrus flavorings were apparently best adapted to the earliest attempts at bottling flavored beverages. Lemon, as previously mentioned, predominated in the bottled drink field until ginger ale was introduced. In these earliest years it was much more difficult to bottle a variety of flavored products because of problems of fermentation and decomposition. The number of different flavors available to early bottlers was considerably limited when compared with those used by the pharmacist or dispenser at the soda fountain. Therefore, these problems of flavor and color seriously restricted the output of early bottled beverages, except for the small beers during the early years of the 19th century. At the same time, however, the soda fountain was using four or five standard flavorings — lemon, vanilla, raspberry, strawberry and ginger.

These earliest flavors were largely manufactured by soaking actual roots and herbs in alcohol to draw out the natural oils. Although many of the methods did not capture the more sophisticated flavors like those currently being manufactured, they did produce a recognizable flavor. Fruit flavors were obtained largely by mashing and mixing fresh fruits with sugar syrups, but because of their perishable nature they were usable only on the day they were prepared. In the 1860s and 1870s the list of available flavors grew longer and included raspberry, currant, cherry, mulberry, strawberry, barberry, lemon, orange, ginger, pineapple and violet flowers. By comparison, the list of the 1890s was literally endless with the addition of almond, asphodel, banana, blackberry, blood orange, calisaya, catawba, checkerberry, cherry, chocolate, coffee, crabapple, grape, phosphate, syrup of violets, walnut cream, wintergreen, and dozens of others (Fig. 37). Tufts, in his catalogue of soda water supplies (ca. 1885), lists other combinations of flavors which could be made, including "Ambrosia," a combination of raspberry, vanilla, and hock wine; "Goldenade," consisting of lemon and egg; "Juliet," which contained ginger, pineapple and grape; "Siberian Flip," an orange and pineapple mixture — the list is literally unending. Keep in mind, however, that these flavors were used only at the soda fountain and, for the time being, were not available to the bottler.

As indicated previously, the first flavorings were, of necessity, derived from freshly crushed fruits. As the soft drink business and the commercial manufacture of flavors increased, each flavor continued to be identified with the name of the fruit from which it was made or which it simulated although no designation was made as to its genuine or synthetic nature. Difficulties occurred when true fruit extracts would ferment in the bottle and the citrus oils underwent an oxidation which resulted in a rancid taste. In 1884, Foote and Jenks were successful in producing the first soluble citrus flavors which were not subject to the rancidity of the older processes; this added greatly to the popularity of citrus drinks as a bottled product.

The smaller bottling plants usually limited their flavors to what was then known as the red, white, black and brown line — strawberry, lemon, sarsaparilla, and cream soda (vanilla). Generally, all bottlers carried a line of ginger ale. The development of a correct color for the various flavors also posed a problem. Early red colorings for drinks such as strawberry were derived from cochineal, carmine or cudbear; most of the brown and black sodas used caramel for coloring which was derived from burnt sugar. At the turn of the century a number of other developments in coloring were made. However, the Federal Pure Food and Drug Act of 1906 placed numerous restrictions on the contents of bottled beverages, and these restrictive limitations forced a close examination of the additives.

Other advances in food chemistry in the latter part of the 19th century yielded artificial fruit flavors of varying types. The production of the first synthetic vanilla in 1872 considerably enhanced the use of this flavor.

UNION EXTRACT WORKS.

Concentrated GINGER ALE EXTRACT Aromatic

4 OUNCES TO 5 GALLONS SYRUP. Price per Gallon, $8.00; per Pint, $1.50.

		PER GALLON.	PER LB.
BIRCH BEER EXTRACT,	(5 ounces to 5 gallons syrup)	$8 00	$1 50
TONIC BEER EXTRACT,	(5 " to 5 " ")	8 00	1 50
ROOT BEER EXTRACT,	(5 " to 5 " ")	8 00	1 50
OTTAWA BEER EXTRACT,	(5 " to 5 " ")	8 00	1 50
PERUVIAN BEER EXTRACT,	(5 " to 5 " ")	8 00	1 50
SPRUCE BEER EXTRACT,	(5 " to 5 " ")	8 00	1 50
NECTAR CREAM BEER EXTRACT,	(5 " to 5 " ")	8 00	1 50
MAPLE BEER EXTRACT,	(8 " to 5 " ")	10 00	2 00

Concentrated LEMON EXTRACT Soluble

5 oz. to 5 Gals. Syrup. Price, per Gal., $8.00; per Lb. $1.50.

		PER GALLON.	PER LB
SARSAPARILLA EXTRACT,	(5 ounces to 5 gallons syrup)	$8 00	$1 50
RASPBERRY EXTRACT,	(5 " to 5 " ")	8 00	1 50
STRAWBERRY EXTRACT,	(5 " to 5 " ")	8 00	1 50
PINEAPPLE EXTRACT,	(5 " to 5 " ")	8 00	1 50
PEAR CIDER EXTRACT,	(5 " to 5 " ")	8 00	1 50
CHAMPAGNE CIDER EXTRACT,	(One pint of which will make 350 gallons cider)	8 00	1 50

LATEST ANGOSTURA SODA EXTRACT NOVELTY

8 Oz. to 5 Gals. Syrup. Price, per Gal., $12.00; per Lb., $2.00.

		PER GALLON.	PER LB.
HOP NECTAR EXTRACT,	(8 ounces to 5 gallons syrup)	$10 00	$2 00
GINGER CHAMPAGNE EXTRACT,	(8 " to 5 " ")	10 00	2 00
CREAM SODA EXTRACT,	(5 " to 5 " ")	12 00	2 00
CHAMPAGNE PUNCH EXTRACT,	(8 " to 5 " ")	12 00	2 00

BARLEY MEAD EXTRACT,

5 Oz. to 5 Gals. Syrup. Price, per Gal., $10.00; per Lb., $2.00.

FULL INSTRUCTIONS WITH EACH ORDER.

VANILLA, ORANGE, ROSE, CHERRY, NECTAR, COFFEE, CHOCOLATE, JAMAICA GINGER EXTRACTS, &C.

All these CONCENTRATED EXTRACTS are SOLUBLE and will remain CLEAR.

RASPBERRY AND STRAWBERRY COLORS. Will not fade nor stain. Price, per Gallon, $12.00; per Lb., $2.00.

ALL KINDS OF ESSENTIAL OILS, COLORINGS, FOAM, ETC.

J. H. MERGENTIME & CO.,

155 CHAMBERS STREET, NEW YORK.

Figure 37. This Union Extract ad is indicative of the wide variety of flavors available in the late 19th century (*N.B.G.* 1886).

One pound of the true vanilla extract cost about three hundred dollars while the artificial flavoring could be marketed for less than ten dollars for a like amount. A variety of flavors seemed to have reached a period of peak popularity during the latter decades of the 1800s and early 1900s but soon lost favor with the public. On the other hand, some, such as the colas, continued in popularity to the present. From 1900 to 1910, a number of phosphate drinks became popular; and there was an early attempt to popularize peach drinks. During the period, 1910 to 1920, grape drinks were the vogue with cherry and lime drinks becoming well known in the latter part of that era. In 1916, "Orange Crush" was among the first of the cloudy orange preparations, but it soon faced competition from several other firms. Orange drinks remain among the most preferred today.

It is worthy of note that, during the early years of the soft drink industry, one of the ever present competitors of the soda pop bottler during the summer months was the watermelon. In the September 1884 issue of the *National Bottlers' Gazette,* a discussion was presented relative to the actual competition which one could expect between a drinker (or as occasionally phrased, a confirmed "soaker") of carbonated beverages and the consumer of the watermelon. Although no one was sufficiently bold to suggest that the watermelon was capable of putting the soda water plant out of business, the relative competition from this particular fruit was discussed with some seriousness. Melon flavors were a relatively late addition to the soda pop line and may have been introduced to counteract this villain. The accepted fact that the watermelon was a competitor of the soft drink is clearly brought out in an early brochure (ca. 1900) distributed by the Cairo Food Products Co. of Cairo, Illinois, for "Mint Cola" (see Chapter IX, Fig. 67).

To arrive at the final product, the flavoring (extract) was mixed with a "simple" syrup solution and carbonated water. The basic formula for manufacture of a simple syrup was 50 pounds of granulated sugar to 4½ gallons of water. The water was heated sufficiently to dissolve the sugar and then filtered while hot through an appropriate fine screen of flannel or felt or a silk bag to remove the unwanted materials. A syrup made of approximately these proportions was of itself a natural preservative. Bottlers using syrup with 6, 8, or 10 pounds of sugar per gallon of water were supposedly faced with such products turning bad.

An additional additive generally used with the syrup extract and coloring was a soda "gum" foam (Fig. 38) which produced the apropriate head or layer of foam in the opened beverage. Although included in most formulae for manufacture of the varying flavors, Tufts was quick to note that foam should not be added to goods for shipment and that too much foam made the beverages ropy. (The term "ropy" was used to describe any off-flavor, taste, or smell which developed in the bottled beverage.)

GINGER ALE

Perhaps the first major flavor to literally revolutionize the soft drink market was that of ginger ale. Its origin is attributed to a Doctor Cantrall of Belfast, Ireland, around 1850. Soda pop bottles with the name Cantrall and Cochran have been found in various parts of the United States where the product was soon sold as a major export item from Ireland. Large quantities of ginger ale were shipped to the United States from several Irish firms, much of it in bottles with only small amounts in barrels or in extract form. Many of the early round-bottomed and pointed-bottomed bottles thus came from a European manufacturer and were shipped to the United States. As a shipment, they provided excellent ballast in the bottom of the sailing vessel.

GEYSER GUM FOAM.

The undersigned has a splendid Formula for making Gum Foam. On receipt of $10.00 I will send

Five Gallons of my Geyser Gum Foam

and enough ingredients to make FIVE GALLONS more, with directions. The ingredients can be had in any drug store. Will send sample pound for fifteen cents. Address,

H. B. BISCHOFF,
COLLINSVILLE, ILL.

FIGURE 38. A necessary additive to all soda waters was a foaming agent (*N.B.G.* 1885).

In the later years of the 1800s, about 3½ million bottles of ginger ale were imported annually with an estimated value of $215,000. There was a duty charged on the imported ginger ale during these years although mineral waters from abroad entered duty free. These lighter types of mineral waters were brought in at the rate of almost two million gallons each year. After the initial development of the ginger ale trade by Irish firms, a number of English manufacturers soon joined the market. A ready acceptance by the American consumer of these varied ginger ales brought about one of the first major huckster operations in selling flavors. Literally every firm which offered a ginger ale extract claimed theirs to be better, of superior flavor, and more concentrated than that of any other competitor. If one were to believe the facetious advertising as implied in Figure 39, one ounce of the extract would make a thousand bottles of ginger ale; competitors claimed that their product would make five thousand bottles per ounce! Ginger ale provided the cartoonist of the day with a wealth of ideas (Fig. 40).

The first soft drink trademark registration was issued in 1871 for "Lemon's Superior Sparkling Ginger Ale." The Clicquot Club Company began the manufacture of ginger ale and other beverages in Millis, Massachusetts, in 1881. Much later, during the 1920s, a new type of ginger ale was developed — the dry ginger ale. In contrast with the sweet golden ginger ale previously manufactured, this newer form was lighter in sugar content and heavier in lemon flavor. Its appearance on the American market came in 1921 when Canada Dry began

FIGURE 39. Competition to sell ginger ale extracts drove various firms to make ludicrous claims for their product. This cartoon parody let the industry laugh at itself (*N.B.G.* 1884).

exporting the product although actually it had been manufactured for many years in parts of Canada. Popularity of dry ginger ale increased rapidly and tended to supersede the older form of ginger ale as a major industry product.

As an interesting sideline, "Goldelle" ginger ale was a franchised extract sold by the Coca-Cola Bottling Company during the first quarter of the twentieth century. The extract to manufacture this drink was available only to bottlers of Coca-Cola and was sold to them as a subsidiary product in their soft drink line. Although reasonably well advertised at the time the drink apparently never attained any substantial portion of the market and was subsequently discontinued by the firm.

ROOT BEER

One of the most effective and profitable enterprises tried by flavor manufacturing and bottling companies was the production of root beer extract for home use. Charles E. Hires of Philadelphia was a pioneer in this venture, beginning with the manufacture of the extract in 1876. Hires did not start bottling his root beer until almost 20 years later. Although the bottled drink soon became very popular, the Hires company continued to produce the extract for fountain and home use. The concentrated extract came in three-ounce bottles with instructions for correct preparation (1 bottle of extract

GINGER ALE—OR WHAT?

TRAMP.—"Say, boss, what was that stuff you gave me yesterday—an acid?"

BARKEEPER.—No; it was ginger ale. Why?"

TRAMP.—"Well, I didn't know; but every time I wiped my mouth I burned a hole in my shirt-sleeve!"

FIGURE 40. Formulations for many beverages were haphazard but probably not this bad (*N.B.G.* 1884).

to 4¾ gallons of water and ½ of a yeast cake) and the bottling procedure.

The use of root beer extracts for preparing the drink at home was popular because the ingredients cost very little (Hires extract sold for twenty-five cents a bottle) and almost any type of bottle could be used to hold the drink as long as it could be sealed tightly. The early small, four-sided extract bottles were embossed with the lettering: HIRES HOUSEHOLD EXTRACT/FOR MAKING ROOTBEER AT HOME/MANUFACTURED BY THE CHARLES E. HIRES CO.,/PHILADELPHIA, PA. U.S.A. The success Hires encountered with extract for making root beer at home soon caught on, and before long several other firms were producing a similar product (Fig. 41). One of the earliest was Bryant's: BRYANT'S ROOT BEER/THIS BOTTLE MAKES FIVE GALLONS/MANUFACTURED BY WILLIAMS, DAVIS, BROOKS & CO./DETROIT, MICH. During the early decades of the 20th century other companies began selling root beer extract in similar amounts and bottles; two of these were the Chas. A. Liemke Co. and Thompson and Taylor's: STAR BRAND EXTRACT FOR HOME USE/MANUFACTURED BY CHAS. A. LIEMKE CO. ST. LOUIS, MO./LEADING BRAND OF THE WORLD FOR MAKING ROOT BEER/GENUINE HAS THIS SIGNATURE *Chas. A. Liemke*, and THOMSON AND TAYLOR'S/ CONCENTRATE FOR MAKING ROOT BEER/ SHAKE THOROUGHLY/IT FOAMS. Hires apparently continued to lead in the production and sale of root beer extract for home use, however; and although there is far less demand for it now than 50 years ago, this extract is still on the market.

Cola

The use of cola-flavored beverages is relatively recent in relationship to other flavors. Preceded in large part by lemon, grape, ginger ale, root beer, and sarsaparilla, cola beverages did not attain appreciable use until approximately 1890; and it was around the turn of the century before they constituted a substantial portion of the soft drink market. Coca-Cola, perhaps the best known of the cola drinks, was one of the very first of this type and quickly dominated the market in the United States. Manufactured from extracts of the cola nut and from the leaves of the coca plant, this beverage opened a new era in American soft drink bottling. By using extracts from the nut of the West African cola tree as a base for the formula, most firms derived a fluid extract by macerating the ground nuts in alcohol in order to dissolve and extract the flavoring components. Although an important part of the cola beverages, the cola nut flavor is bitter in concentrated doses. The balance, and perhaps more important portions of the flavor, comes from other fruit and spice elements which are mixed with the cola nut extract.

Although generally accepted as a product of American origin, the cola drink was known in England and Scotland in the 1880s and was quite popular in the latter country long before it attained any real favor in the United States. It is reputed that African natives have made cola drinks for many centuries and used them as a cure for intoxication among other things.

The addition of extracts from the coca plant are not always considered necessary to the manufacturer of cola beverages and are omitted in many formulae. The earliest Coca-Cola syrup contained cocaine, a direct extract from the coca leaf, but during the early years of the twentieth century the formula was refined to remove the cocaine from the finished syrup.

Despite its major influence in the American market today, the cola drink was not well established in the early days of bottling. A distributor would often give away a free bottle of Coca-Cola if the purchaser bought a case of another flavored beverage; this was done in hopes that the cola flavor would be accepted and subsequently purchased by the consumer. Although acceptance was relatively slow to come, within a few years colas became one of the most popular forms of soft drinks on the American market.

FIGURE 41. Extracts of root beer (left and right) and juniper-ade (center) were commonly sold during the late 19th and early 20th centuries. These products had lost their popularity by the mid-1900s.

Chapter VI
THE BOTTLED PRODUCT

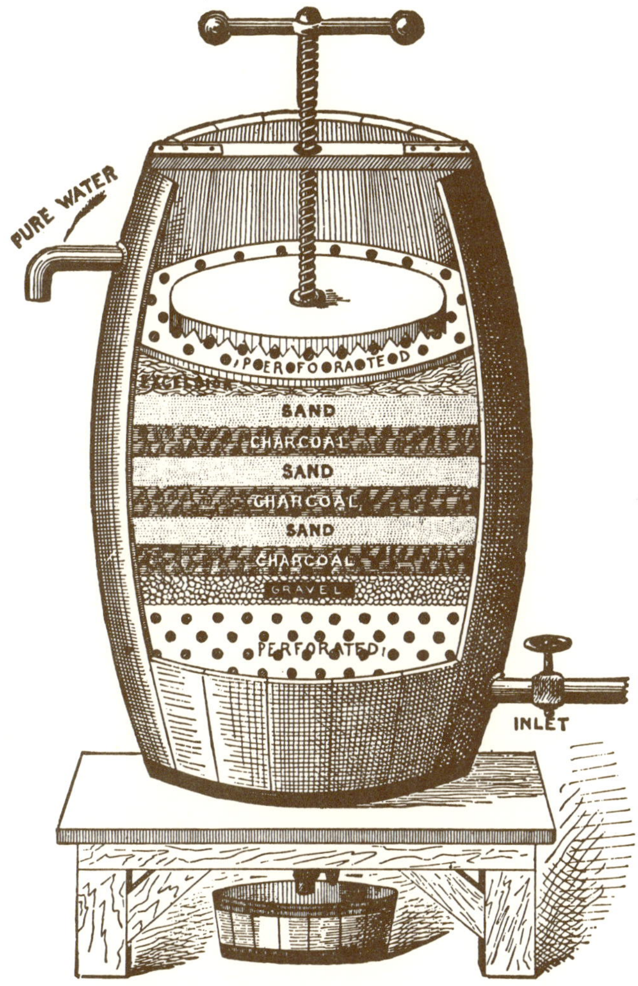

Figure 42. Diagrammatic illustration of a barrel water filter recommended for early bottling plants (Tufts, 1885?).

Despite the long period of development and manufacture of soda water, beginning around 1800, only 64 bottling plants existed in the United States by 1850, these being concentrated largely in the major cities of the East. Fifty years later the state of Illinois alone contained 182 plants, with nearly one-fourth of these in Chicago. At the same time there were thousands of soda fountains selling soda and mineral water by the glass. The soda fountain was the principal competitor of the soda water bottler and, as late as the 1880s and 1890s, over 70 percent of the soft drinks were sold at the fountain. With the improvement of transportation facilities after 1900, bottled drinks began to monopolize a greater proportion of the market. Direct home delivery was one of the major reasons for increased consumption of bottled drinks while distribution and sales through local grocery stores formed an extremely small portion of the annual market.

By 1880, 512 bottling plants were in operation in the United States; in 1890, 1,377; and by 1900, 2,763. Most of the earliest equipment was manufactured primarily for use in the soda fountain, and it was only after 1850 that types of machines designed specifically for bottling plants began to appear in quantity. Since the early bottler was neither hindered by water pollution to the extent currently prevalent nor harassed by public health authorities, he was able to obtain a "pure" water with a relatively simple filter system. Figure 42 shows a cross-sectional view of one of these early filters.

A good whiskey barrel with a volume of about 30 gallons was considered an ideal container. A hole was bored near the bottom and connected to an inlet pipe through which the water was pumped into the barrel; a second hole was bored near the top where filtered water was removed. The actual filtering was done through several porous layers in the middle of the barrel. The filter usually consisted of two to several alternating layers, six to eight inches thick, of coarse ground charcoal and coarse glass sand. These two substances, if thoroughly cleaned regularly or replaced periodically with fresh materials, offered sufficient filtration to remove the gross impurities from the water and yielded a satisfactory level of cleanliness. In some cases a discharge-cock in the bottom of the barrel was used to drain off the accumulated sediment. The charcoal and sand cost about five dollars per barrel, a quantity sufficient to last a year if properly maintained. These filters were easily renewed by a thorough washing in clean water to remove the accumulated impurities which they had subsequently removed from the water.

The clean, filtered water was then carbonated in one of the devices previously described. The two basic systems used for the manufacture of carbonated water were known as the *continuous system* (basically an English or European form) and the *intermittent system* (basically an American derivative). In the continuous system, the CO_2 and water were continuously pumped under pressure into a tank from which the carbonated water was drawn (Fig. 43). Because the sources CO_2 and water were separate from this tank, there was no need to intermittently stop and refill it. However, in the case of the American intermittent system (Fig. 43), individual tanks were filled with water and CO_2 which were agitated or mixed by various means and then used to fill the bottles. Each fountain was

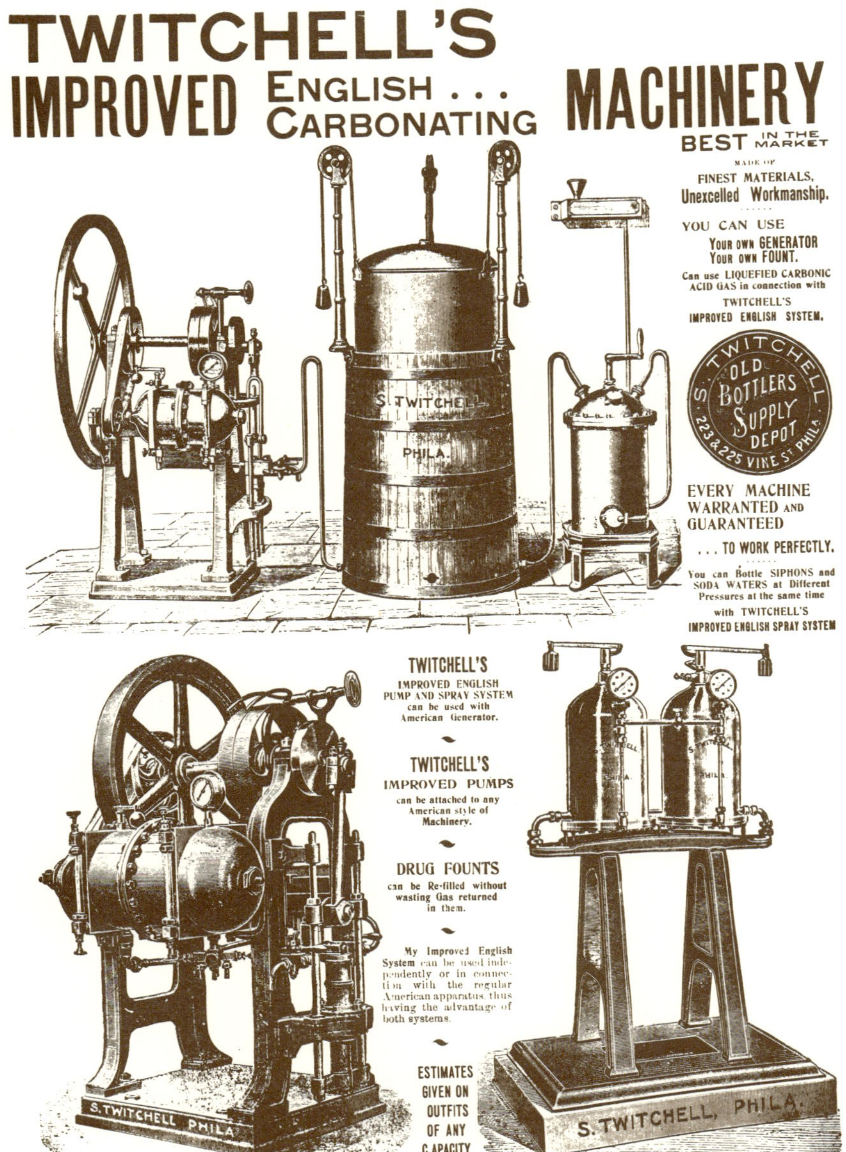

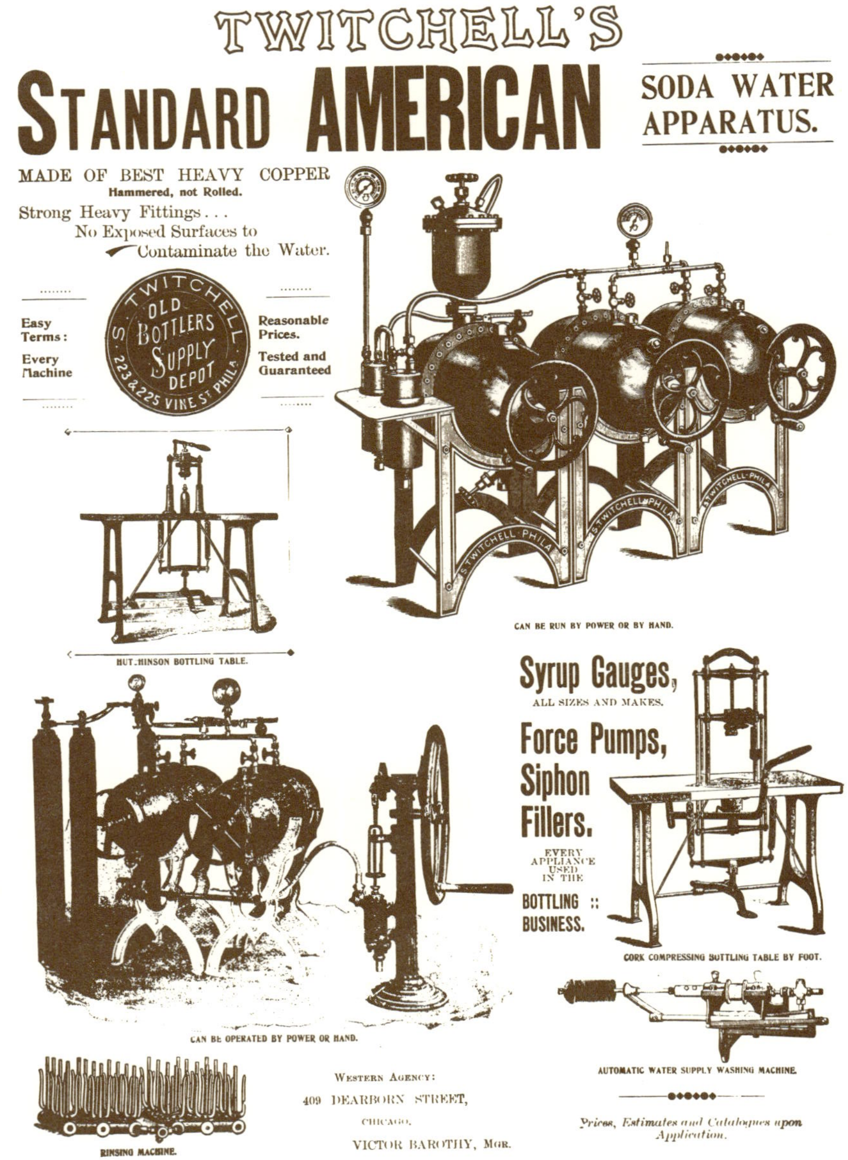

FIGURE 43. The English and American carbonating machinery provided two methods of solving the common problem of making carbonated water (*N.B.G.* 1896).

used until exhausted, then replaced by another full container while the empty one was being refilled with water and gas. An important modification of the intermittent system was developed after a few years of trial and error; a large number of individual fountains were connected in a series so that a more or less continuous supply of carbonated water was obtained by the simple shifting of open valves. For all practical purposes this made the system substantially like its European counterpart.

The filling and capping of bottles was at best a complex and difficult task. By means of a funnel and dipper, each bottle was first given an appropriate amount of flavored syrup; then it was carried over to the source of carbonated water. The procedure of filling the bottle was referred to as "knee" bottling because it depended on the operator's lifting and lowering his knee to maintain the fit of the bottle against the spigot or filling tube. The operator wore a large leather apron, a leather shield or thick glove to protect his hand, a wire mask or goggles to protect his eyes, a knee board, and had at hand a wooden mallet to drive the corks. The bottler opened the discharge valve after placing the mouth of the bottle to be filled against the spigot. A suitable leather packing around the spigot served to make a tight joint between it and the bottle, thus preventing any loss of carbonation. Pressure was maintained by the upward force of the operator's knees which held the bottle against this leather packing. As the bottle began to fill, pressure would develop inside the bottle from air which could not escape until it reached a point at which it equaled the pressure of the liquid coming from the fountain. At this point it was necessary to ease the neck of the bottle off the spigot to release this resisting pressure. More fluid was then allowed to enter the bottle, and this was followed by a subsequent release-fill-release procedure until the bottle was full.

The release of the air confined in the bottle was a process known as "snifting," and it had to be done in such a way as to eliminate the air without releasing the gas contained in the carbonated beverage. Therefore, it was a skill known only to the better men in the bottling business; if improperly done, the result was a flat, uncarbonated beverage. The second major problem occurred when the bottle was filled and it was time to insert the cork. The bottler quickly released the bottle from the spigot, pushed the cork into the neck, and drove it in with his mallet. He then would hand the bottle to an assistant who tied or wired the cork to secure it tightly in the bottle. Again, if this latter operation was improperly done, the best carbonated beverage would be ruined. In fact, the bottling procedure was especially dangerous because bottles frequently burst under the pressure of the carbonation. At best it was an awkward procedure. Continual attempts to improve the method did much to affect the subsequent development of more efficient bottling apparatuses.

Notable improvements in efficiency and safety were made during the 1850s and 1860s when some of the first bottling tables were developed and put into practice. These first mechanical bottlers were relatively simple affairs and did not perform all the tasks subsequently incorporated into the more sophisticated models that evolved during the latter part of the 19th century (Fig. 44). These earliest bottling tables combined a mechanical device for insertion of the cork with the valve which controlled the flow of carbonation. Pressure exerted on a foot treddle would lower the filling head of the fountain; when the bottle was filled, a hand lever activated a piston drive which pushed the cork into the bottle neck. A vent valve was also added to the filling head which provided automatic snifting to relieve the pressure as the bottle filled. Additional safety improvements included the placement of a screen between the operator and his work for protection against injury from bursting bottles.

A subsequent refinement of this basic table was made by the Matthews firm in the late 1860s and involved the application of counter-pressure in the filling operation. In this device the snifting valve, which had been used in some of the earlier bottling tables, was connected in such a fashion that the pressure released by it was directed back into the fountain that was being emptied. This system permitted a uniform pressure to be maintained in both the container being emptied and the container being filled. One of its advantages was that "the uniformity of pressure obtained" would prevent foaming as well as the loss of gas ejected from the bottle in the filling operation. This basic procedure is the same as that utilized in many of the small hand pumps attached to beer kegs that are sold today. If the prevention of excess foaming was no more efficient in the earlier days of the soda water industry than is often the case at a modern backyard party, it must have been disasterous!

All of these aforementioned table machines required prior filling of the bottle with the appropriate amount of syrup by using a dipper and funnel. However, in 1861, Henry W. Putnam manufactured a highly successful syrup gauge which could be accurately preset to dispense a given amount of syrup. Initially devised as a separate addition to the filling head of the bottling table, it was rapidly adapted to the actual filling of each bottle. Although quickly copied by other firms with only minor modifications, each claiming superior performance, the Putnam syrup gauge was the one widely used by virtually all firms and by nearly all bottlers. The development of this gauge and its incorporation into the bottling machine essentially eliminated a person from the bottling crew. One individual was now able to both fill the bottles with syrup and liquid and do the initial sealing while a second person tied the seal and placed the bottle in its storage container.

Bottles employing an internal closure, such as the type made for the Matthews Gravitating Stopper, the Codd style that used a marble in the neck, and others using a floating ball, were largely developed in the 1860s and 1870s. These bottles first received syrup while in an upright position; they were then inverted on the filling head to complete the filling with water and/or carbonation and to seal them. Later modifications in the machine allowed total filling in an upright position; inversion of the bottle was necessary only to drop the closure in place. Despite a reasonable amount of promotion, these internal-stoppered bottles were seldom used in the American market. The subsequent development of the Hutchinson Stopper in 1879 literally eliminated them from the competitive market. The Hutchinson Stopper was not considered an intern-

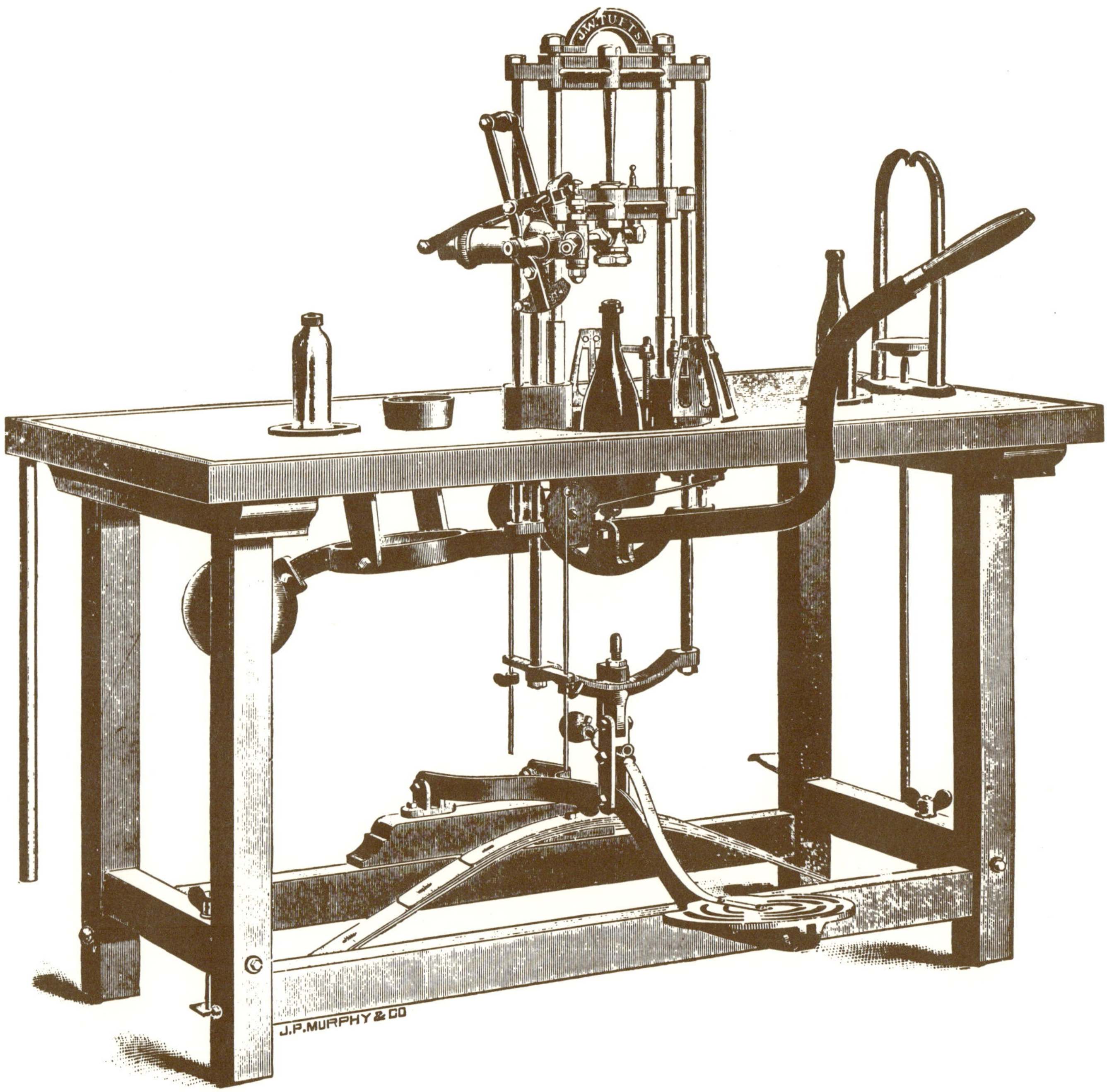

Tufts' Improved Bottling Table.

(PATENTED.)

NEW STYLE FOR 1888.

FIGURE 44. The foot-operated bottling table was in common use around the turn of the century (Tufts, 1885?).

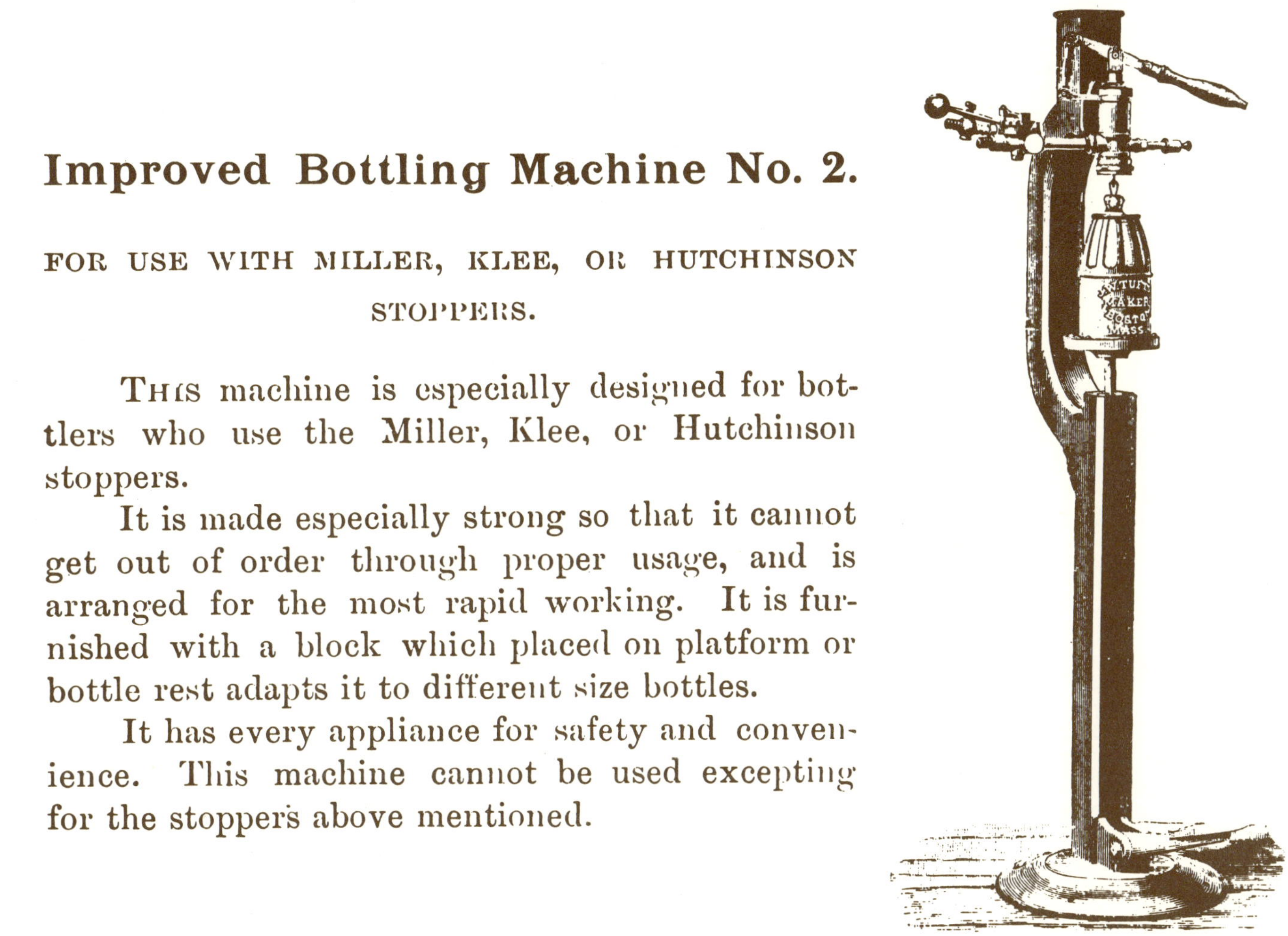

Improved Bottling Machine No. 2.

FOR USE WITH MILLER, KLEE, OR HUTCHINSON STOPPERS.

THIS machine is especially designed for bottlers who use the Miller, Klee, or Hutchinson stoppers.

It is made especially strong so that it cannot get out of order through proper usage, and is arranged for the most rapid working. It is furnished with a block which placed on platform or bottle rest adapts it to different size bottles.

It has every appliance for safety and convenience. This machine cannot be used excepting for the stoppers above mentioned.

FIGURE 45. The hand-filling and sealing operation of the Hutchinson-style soda bottle could be done with "safety and convenience" on machines such as this (Tufts, 1885?).

al closure since the bottle could be sealed while in the upright position (Fig. 45).

Syphon-filling machines were also available and were rather simple devices. They consisted of a bottle holder and a protective screen to cover the bottle while filling and a foot-treddle mechanism which moved the bottle upward to the filling head where the carbonated water was drawn from a fountain. When syrups were used in the syphons, they were added separately by use of a syrup pump.

The development of the crown closure in 1892 revolutionized the bottle-filling and sealing operation as well as the equipment necessary for performing these operations (Fig. 46). The earliest crown fillers for machines were very similar to models used by Painter for the closure of his Bottle Seal.

An early description of the crown seal machine included the following points: The crowns (or caps) were contained in a cup in front of the machine. Two bottles were always in use, moving from left to right across the table of the machine. In operation, the bottle on the right under the head of the machine received the carbonated water while simultaneously the one on the left received its measured quantity of syrup. The bottles then moved to the right, allowing the bottle containing syrup to receive its allotment of carbonated water. When the bottle was filled, a quick step on a foot treddle crowned the bottle. This type of syruper-filler-crowner was operated entirely by foot treddles and hand levers. Adaptations through power sources came only in later models. Automatic crown-feed devices and power machines to complete all the operations automatically were not developed until after the turn of the century. Despite the many advantages of the crown filler, conversion to it was relatively slow, and many bottlers continued to use their older styles of bottles, closures, and bottling tables well into the 1900s.

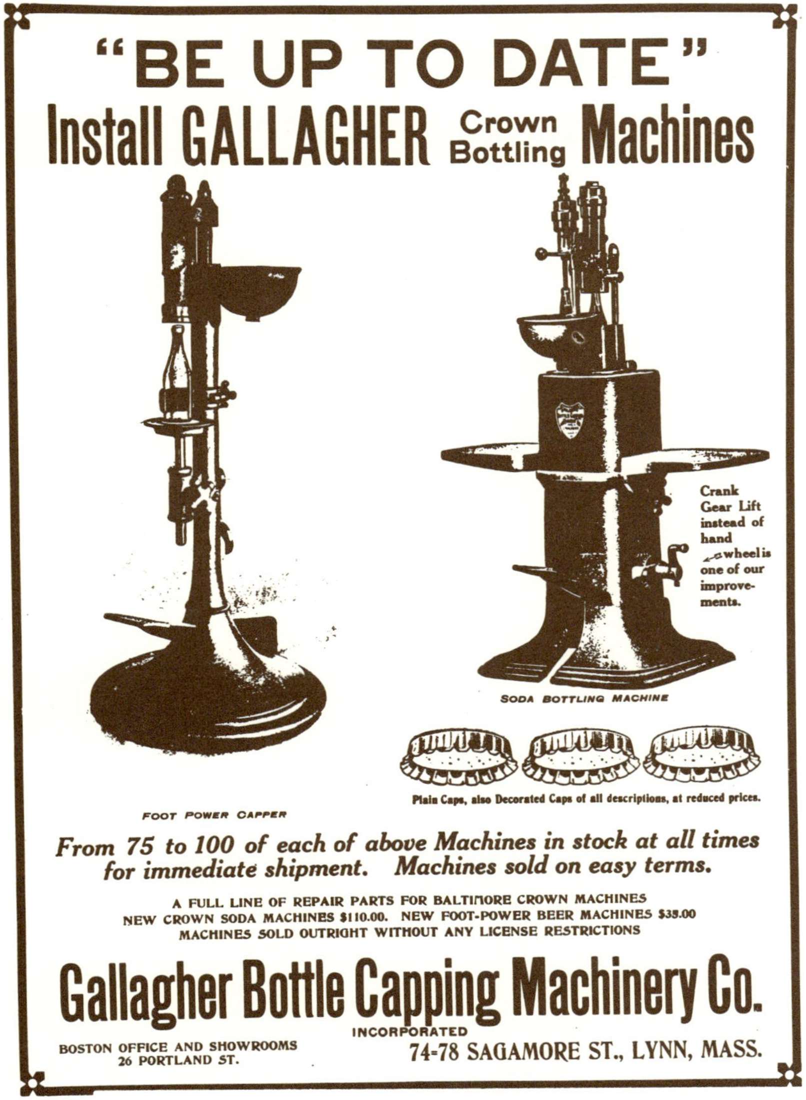

Figure 46. With adoption of the crown stopper, machinery to fill and seal bottles became increasingly more complicated and mechanized as these 1910 ads from the *N.B.G.* indicate.

Chapter VII
PROCEDURES AND PROBLEMS WITH CLEANLINESS

The job of thoroughly cleaning bottles between uses appears to be a relatively late-blooming art within the soda water industry. As late as 1876, the firm of John Matthews offered a new cleaning device, known as the Matthews Bottle Washing Machine, that provided a single jet of water with a pressure adjustment which would cleanse a presoaked bottle, removing any inside accumulations of dirt. Matthews described his apparatus thusly, "When properly adjusted, the slightest pressure of the bottle mouth on the lever will inject a bottle with water sufficient to cleanse it." Thus, the entire cleaning operation consisted of soaking the bottles for a period of time to loosen the dirt, then rinsing with clear water.

In a bit of correspondence to the editor of the *National Bottlers' Gazette* in 1896, a bottler who signed his letter "Progressive" described the conditions under which he and a partner bottled soft drinks. The "Progressive" partner was lamenting the poor quality control of his partner and wanted all to know the extremes to which some persons resorted in order to save a penny. He describes his partner's work habits thusly: "The bottling table had a coating of dried syrup on it 1/4 inch thick. The floor was muddy all the time and was constantly used by this man as a cuspidor. He wouldn't rinse a bottle in clean water, would soak 3 or 4 lots of bottles in the same water, set the bottles of soda in which flys or sediment were found to one side to give away to friends or boys who run errands, or would put them in cases for country store trade saying that 'They wouldn't know the difference out there' or 'It's good enough for Welshmen or Swedes.' " "Progressive" was most pleased that he was able to divest himself of his partner's unworthy assistance and was advertising in search of a clean young man with liberal ideas about business matters to enter the business with him. A cartoon (Fig. 47) appearing in the *National Bottlers' Gazette* over a decade earlier typified the consumer's plight.

TOO PARTICULAR.

Boy.—"I say, mister, I have dirt enough without drinking it in my soda."

Barkeeper.—"Well, what of it? It don't seem to hurt the Bottler, and I reckon you can stand it. You're too particular for the kind of stuff I sell."

Figure 47. Cleanliness was a constant problem for the bottler (*N.B.G.* 1884).

One of the more or less constant hazards of the pop bottler was the tendency of his beverages to develop "ropiness." Since ropiness could result from contamination by almost any form of "dirt," organic or inorganic, and literally during any step of the preparatory or bottling phases, cleanliness was the byword among the more professional bottlers. As implied by the humorous 1884 ad of Bowker (Fig. 48), numerous hazards existed. One bottler went so far as to say that the action of his pump, strictly a device for the transport of water or other liquids from one point to another, led to the elimination of ropiness in his beverages. This was all rationalized in an extremely circuitous argument which was summed up with the implication that he was able to thoroughly clean his entire bottling establishment from top to bottom and rinse it by means of his pump. Therefore, the maintenance of a clean establishment—which was facilitated by use of the pump—yielded a better beverage.

For a period of about 20 years, beginning in the late 1870s and lasting into the 1890s, numerous devices for the improved cleaning of bottles were perfected. The Triumph Bottle Washing Machine (Fig. 49) represented a style of soaker that was adopted by numerous firms although with various modifications. Patented initially around 1880 and used by European companies, the machine consisted of a large wheel with rows of openings for insertion of bottles that continuously revolved through a soaking tank containing wash water. By means of attached brushes and rinsing tubes, it automatically scoured and rinsed the interiors of the bottles. The Triumph machine never gained great favor in the American market, however, probably because of its large bulk and general unavailability, al-

CHEATEM QUICK & CO., Bottlers' Supplies.

The largest establishment in the business on earth or in the moon, covering 40 square acres; main building 17,000 feet long, with two wings 18,000 feet each, 36 stories high. We keep 4,000 drummers, any one of whom can go over North America and some parts of Delaware in fifteen minutes. Special attention is called to our Machinery Department.—Our main shaft is 8 feet in diameter, and makes 60,000 revolutions a minute. We turn out 400 sheet-iron generators every hour, copper lined, tin lined, brass lined, paper lined, or any other lined. This department is under the charge of Mr. Grab Mortgage. Old generators painted over and sold for new; old apparatus taken for nothing, painted over and sold for new, on the instalment plan.

Drummer for Cheatem Quick & Co.

Our Repairing Department is under the charge of Mr. Junk, where all work is done inside of five years at the low price of $50 per hour (not including stock).

Chemical Department. This is a large feature of our immense business, and is in charge of Prof. All Fool, who has had 200 years of practical experience in pumping, mixing and compounding. (See illustration.) One special feature of this department is our filter. For fine work the old hat filter is best adapted; for common use—water, syrups, &c.—the coal sieve is recommended, it is best adapted for new beginners, as it filters 200 gallons a minute. Mud is made perfectly transparent in four seconds; "nothing settles" after using this filter.

Extracts. We have every variety of extracts made by our assistant foreman, Mr. Printer. Imitation Birch, Soap Gum, Soluble Turpentine, Soluble Cayenne Pepper, Soluble Lard Oil for making Tonic Beer, &c.; also Compound Solid Extract of Ginger and Cayenne Pepper, one pound will make 6,000 gallons of Ginger Ale, by filtering out all the Ginger it becomes perfectly clear.

Coloring. This is the boss. For best color use our *Coal Tar Brand*, one drop colors 16 barrels. It has the "heft," 100 pounds to the gallon, and will mix with mud, lard oil, or benzine, and make them come as clear as crystal. Price 25 cents per gallon, if the customer buys a sheet iron generator, and $5.00 without.

Red Color. This is made from pure arsenic and "never settles," one drop colors three barrels of liquid, drives horse and wagon and paints the town red. Sold only to new beginners.

Soap Foam. This we prepare from the original formula obtained from Mr. Stealem, who obtained it from Noah's patent in the year 17. It is the only genuine article in the world made from the original. Furnished free to new beginners if they buy other goods.

Essential Oils. This is our great forte, and is in charge of Mr. De Swindler. We manufacture them all from the original seeds; they are all new crop, "hand pressed," and hen pecked; specific gravity .2000. They never "rile" anything and mix with water perfectly clear. One ounce to the gallon.

Terms. We sell all of our goods on the instalment plan, and prefer to sell them through our drummers. We do not care for direct orders, as our drummer knows better what the customers want than they do themselves, and will send in all cases a full line of goods from two to four years in advance. Our drummer will call every fifteen minutes and furnish refreshments free. In order to please every one our price list is printed on India rubber; prices are always made to suit the customer. Parties buying any of our sheet-iron generator machinery, or having any repairing done, the extracts will be sold very low. To parties who contemplate buying an outfit the extracts are free. The great cheapness of our goods consists in the small quantity used, "one ounce to the gallon," from that down to the millionth of a drop, as in our red paint color. New bottlers, and those who have a poor trade, will please notice this fact. No charge for packages, this is all taken out of the goods when they are put up. Accounts can run seven years at 50 per cent interest. Notes will be required which mature once in three months, and will be paid to our attorney, Mr. Suem Quick.

P. S. Any new flavor, extract or compound will be furnished as soon as the label can be printed, at half the price of the genuine. All of our drummers carry receipts which will be given away free.

Special Notice. Our new Kerosene flavor will soon be put upon the market, which will supersede all other flavors, on account of the small quantity required, *one drop* to the *gallon*. In addition to the great variety of articles we now furnish to the trade, we shall increase our list so as to furnish any article that bottlers may require, including bottles, patent stoppers, clothing, household furniture, cooking stoves, hosiery, horses and wagons, drivers, harnesses, stationery, cork screws, cayenne pepper, books, coffins and groceries.

The list will be increased from time to time.

For further information address Cheatem Quick & Co., New York, Boston, Chicago, Philadelphia and St. Louis.

For all standard reliable preparations used in the manufacture of soda waters, cider, wine or vinegar, with the latest improvements, address **Dr. H. L. BOWKER & CO., 295 & 297 Franklin St., Boston,** and you will receive circular with full information free of charge.

FIGURE 48. The Bowker ads emphasized all the pitfalls and tricks of the trade to which a bottler might succumb. With tongue in cheek, they presented humorous ads parodying the entire bottling trade (*N.B.G* 1884).

THE

"Triumph" Bottle Washing

MACHINE,

AS USED IN LONDON AND PARIS.

The attendance of a boy or girl only is required to clean

150 DOZEN PER HOUR.

The cleansing of the bottles is effected by brushes or rubbers and rinsing tubes, which **automatically** operate upon **SIX BOTTLES AT ONCE, without their being handled by the attendant.** The above sketch represents a machine holding 14 dozens—six in a row and 28 round—and occupying about 6 feet by 5, but machines of a larger or smaller capacity can be made.

PROTECTED BY FOUR PATENTS.

For Full Description and Prices Address

THE TRIUMPH BOTTLE WASHING MACHINE CO.,

70 High St., BOSTON, MASS.

FIGURE 49. The Triumph Bottle Washing Machine, while a major advance for its time, did little more than soak and rinse the bottles in a water solution (*N.B.G.* 1886).

though similar soakers based on this continuous-wheel principle were used by smaller firms. These wheels operated in such a manner that, as a bottle was added on one side, the additional weight of another bottle would cause the wheel to rotate; the soaked and partially cleaned bottles were then removed from the other side of the wheel.

A number of firms developed small washing devices which became popular in the small-town bottling plants. The Lightning Bottle Washer, as well as several competitors of similar construction, consisted of a single rotating brush and rinsing device (Fig. 50). Bottles were singly hand-fed and held on the machine until thoroughly washed; they were then removed for a final rinsing and placed on a drying rack or draining truck. These machines were largely operated by footpower from a treddle although occasionally adapted to steampower. They were not readily adapted to large operations because of their obvious limitation of washing only a single bottle at a time. However, because of their simplicity, they were used in a majority of the bottling firms from the late 1870s well into the 1890s.

Among the variations to be found in these earliest single-bottle washers was the type of brush used in the scouring operation. The Lightning washer was advertised as having a spring brush made of rubber and flexible steel while other washers had corrugated rubber or bristle brushes. The next logical step in washing was the development of parallel rows of brushes where 2, 6, 10, 20 or more bottles could be washed simultaneously.

Another of the early styles of bottle washer was the shotting machine. These devices were simply another means of scouring and loosening the inside accumulations in bottles and were used, as was the Lightning-style machine, in conjunction with separate soaking and rinsing operations. Shotting machines were generally operated by footpower or steampower with fine steel shot as the scouring medium. Reputedly they could wash all styles of bottles and thoroughly clean the stoppers and necks simultaneously. The basic machine was simply a cradle which held the bottles filled with shot and, by rocking them gently back and forth, scoured the insides clean. The Donally Shotting Bottle Washer was described by the manufacturer as using either steel or porcelain shot. The machine automatically injected a pound and a half of shot into each bottle which was "shaken more thoroughly in the bottle than could be done by hand." After the shaking operation, the bottle was held in a vertical position and the shot and water drained; as a final step a stream of fresh water was forced into the bottle to wash out any remaining particles or sediment.

One of the disadvantages of shotting was that frequently the shot could not be used a second time. Shot thrown into a greasy bottle became so coated with oily matter that it caused more harm by contaminating addi-

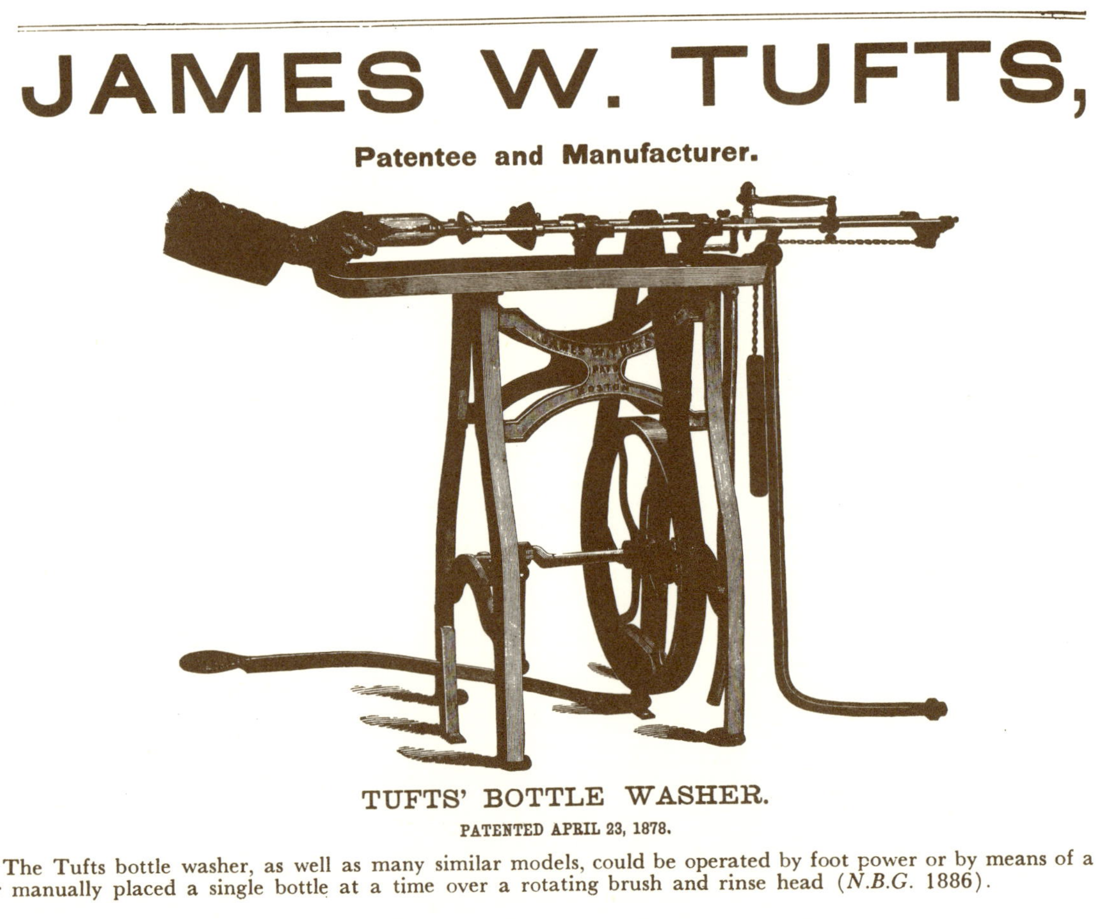

FIGURE 50. The Tufts bottle washer, as well as many similar models, could be operated by foot power or by means of a steam engine. The operator manually placed a single bottle at a time over a rotating brush and rinse head (*N.B.G.* 1886).

tional bottles than it did good in helping to clean them.

A device similar to the shotting machine, called an English Pebble Machine, followed the same principles but used small pebbles as a means of scouring the bottles. By handcranking the machines, two counterbalanced bottle-holding devices gently shook the bottles till thoroughly cleansed. There has been seemingly little change in principle from these earliest hand-powered mechanisms to the more sophisticated machinery of today other than an elimination of manpower and an increase in capacity. Today bottles are loaded in mass and run through machines which soak, scrub, and rinse automatically.

Various suggestions were printed in the *National Bottlers' Gazette* (1896) for removal of crusts or impurities which resisted the more standard cleaning procedures:

1. Soak them in a solution of permanganate of potassium.
2. Rinse the bottles out with a solution of equal parts of muriatic (hydrochloric) acid and water.
3. Allow the bottles to lie in a solution of chloride of lime and water, in proportion of one ounce of lime to two pints of water, for 3 or 4 days.
4. Soak in a mixture of potassium bichromate and sulfuric acid.
5. Strong sulfuric acid put in the bottles, corked, and allowed to stand a day or two should remove the strongest crust.

As the American consumer became more appreciative of soft drinks and consumption levels increased substantially, one of the outstanding advertising features (especially by the major proprietary brands) was based on the fact that the product was contained in sterilized bottles. Prior to the turn of the century, the actual sterilization of bottles was seldom done in the cleaning process. Figure 51, showing bottle-washing machines of 1910 and 1915, illustrates the rather noticeable increase in size which occurred with the more modern types, a factor that allowed the bottler to increase his production and provide a satisfactory product.

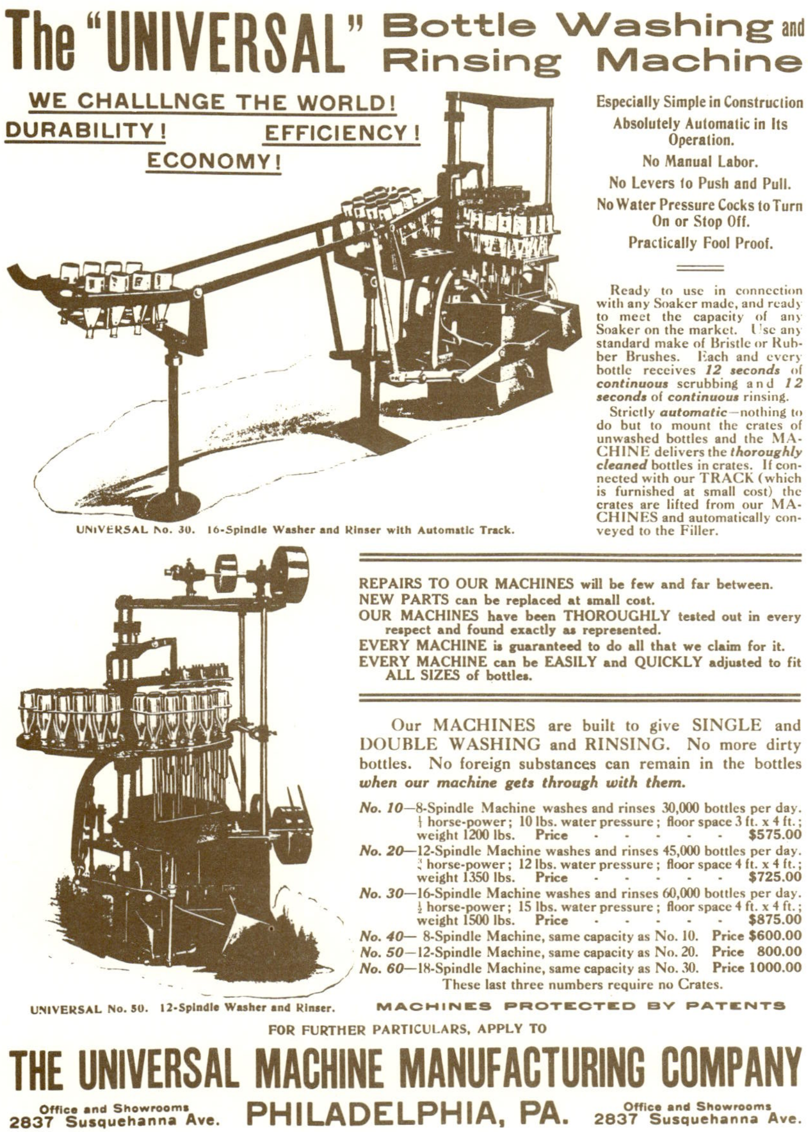

FIGURE 51. The "Universal" Bottle Washing Machine of 1910 (*N.B.G.*) and the "Loew Junior" of 1916 show the rapid advance in mechanization of bottle washing in the early 20th century.

Chapter VIII
LABELING, ADVERTISING AND DISTRIBUTING THE PRODUCT

The value of good advertising has long been shown to be a profitable investment in any business. The phenomenal success of the Coca-Cola Company, which in a large part resulted from an early and extensive advertising program, provides one of the best examples. Other major soft drink companies, for example, Hires Root Beer, also learned in those early years of operation that extensive advertising programs eventually paid off in substantially increased sales of their product. In addition to some of the more generally well-known advertising methods and programs, several examples of which are cited in this and other chapters, a number of suggestions were made by James W. Tufts in his 1885 catalog of bottling supplies. He suggested that the bottler maintain an attractive show window to display his product; the bottling apparatus should be set up in the front of the shop, as near as practicable to a door or window, so it could be easily seen from the street. In this way the operation of the machinery would attract the public's eye.

It seems Tufts had considerable foresight as this type of business philosophy has been retained until recent times; in fact, the actual bottling and washing apparatus of most major firms was placed behind large plate glass windows where the public could readily observe the operation. Appearance to the public was deemed an important feature. "Nothing advertises a bottling business more than neat wagons, horses, and men" (Tufts, 1885).

Other important tips included the following: 1. Do not spend money at bars to advertise your business. It is very expensive advertising and, besides, it is unprofitable. 2. Beverages should be sold on their merits, and treating (giving free samples) should be left to the brewers and distillers who originated the practice. 3. Put your name prominently and permanently on every article used in your business.

It is this last piece of advice that has resulted in the preservation of so many identifiable artifacts which have become the core of a highly collectible group of memorabilia. As the beginning collector soon learns, the name (as well as the actual or inferred date of use or manufacture) found on the artifact is the key to its relative interest and value to other collectors. Local materials, e.g., bottles formerly used in a given community, are more collectible in that community than they are in other locales. Exceptions to this rule include items which were distributed on a national basis and those which, because of their extreme rarity, are of general interest to all collectors (such as the cobalt soda bottles of Ryan of Savannah, Georgia, and Lomax of Chicago). It is, in the final judgment, the identification of objects and advertisements of the product which provide interest and special satisfaction to the collector.

Labels

Identification of bottles used by the early soda water manufacturer was facilitated by molding his name and/or the company's name into the glass container, but the name of the product (contents) was seldom embossed. Only by knowing the manufacturer — and often by simply guessing from the color of the liquid — could one surmise the contents of a bottle. The additional use of a paper label, although widely used since early days, was far from uniform. Riley (1958) implies that the paper label was seldom used except for the more popular items, even in the early years of the 20th century. He quotes an old-time bottler as stating, "I recall that my entrance into the industry was in 1907 when it was my job to use a flour paste which, by means of a brush, was applied to a wide 5-foot-long board on which the labels were placed after they were bent to a convenient shape for handling. From this board they were then placed on each bottle by hand." So-called automatic labeling devices were available by the turn of the century although they offered little improvement over the hand-labeling procedure. Some of these earliest labelers did little more than apply the paste to the label mechanically, it still being necessary for the operator to press the label to the bottle by hand.

Automation was sure to follow, however, and with an increasing demand on bottlers to produce greater volumes of their product, it seems new devices were perfected in time to solve the current problems. During the earliest years of the 1900s, several models of hand-fed labeling machines were produced, some of which could process as many as 20 bottles per minute. By the 1920s, nearly total automation had arrived in this segment of the industry and labelers capable of marking 80 to 100 bottles per minute were on the market.

The paper label, however, was not destined to be the epitome in bottle marking. The applied color label, a fired-on, vitreous color applique, was developed in 1934 (Fig. 31). With this new process, permanent surface decoration and identification could be applied to the bottle. Performing essentially the same functions as the paper label, it would also persist through constant reuse and cleaning. This caused two major changes in the industry: first, a gradual elimination of paper labels except for use on bottles to identify the various flavors sold by the same manufacturer; second, the demise of many of the proprietary bottles which required special designs and molds for their manufacture. Through the liberal use of design and color with this baked-on label, one could effectively separate and characterize his bottle without need of a unique shape.

FIGURE 52. Paper labels such as these were used extensively by private bottlers during the 1920s and 1930s. While the label on the right was used for a locally produced product, the Dr. Pepper label was a part of a franchise arrangement.

The evasive paper label has caused no end of trouble to collectors who always attempt to find the earliest bottles of the various proprietary lines. Virtually all of the proprietary lines which survived for many years, several of which are still active today, for example, Coca-Cola, Hires, and Dr. Pepper (Fig. 52), initially sold their product as an extract or flavoring and did not bottle it in the distinct, characteristically identifiable bottles so well known today. As seen in Figure 53, which illustrates the "Bull Ginger Ale" label, bottlers were offered the option of buying the extract to manufacture ginger ale following the Bull Ginger Ale formulation; with this option they could also buy the trademarked label to which their name could be added. Thus, the chance of finding a recognizable bottle of Bull Ginger Ale is slim since bottles possessing descriptive embossing identifying the product were probably never produced. Perhaps hundreds of bottlers used this formulation and bottled this popular brand in their own bottles, but the use of perishable paper labels leaves room for only guessing.

Over 7,000,000 **Bottles (Seven Millions) of R. & R. Bull Ginger Ale were bottled in 1890. There is no compulsion to use our label if you prefer your own.**

THE BULL GINGER ALE EXTRACT, manufactured from the personal experience and study of our Mr. Max Rosenheim in the City of Belfast, Ireland, is gaining daily in favor with the bottlers of the United States. They have found it in every respect an article that enables them to make a Ginger Ale equal to the best imported brands in the most simple and quickest way. While the Bull Ginger Ale Extract is of the greatest strength, requiring only 1 oz. Extract to 1 gall. of Sirup or 4 ozs. Extract to 5 galls. of Sirup, it mixes perfectly clear with sirup and water without any filtration or manipulation whatever, remaining sound, clear and perfect until consumed. ***At the request of many esteemed customers who bottle the "Bull Ginger Ale," we have now gotten up and secured by law the above "Bull Ginger Ale Label,"*** which will be supplied by us at cost price with the Extract. There is, however, no compulsion whatsoever on our part to use this label if bottlers prefer their own; we merely offer it, believing it to be an advantage and advertisement to the bottlers. Bottlers wishing to secure a steady increasing paying trade should not miss to try this extract.

Price of Bull Ginger Ale Extract, per gall., $8.00; 5 gallon lots, per gall., $7.50; 10 to 40 gallons, per gall., $7.00.

Directions for use with each order.

OUR TESTIMONIAL SPACE

KINGSTON, JAMAICA, April 6, 1891.

THE R. & R. CHEMICAL WORKS, 44 Trinity Place, New York.

Dear Sirs:—We are more than pleased with your Essences. Shipment per Steamer Crescent received in A 1 order. "Bull Ginger Ale" Extract highly satisfactory; we will use no other, of this you may be assured; it beats ——————, of Hull; ——————, of New York; ——————, of London; in fact, *is best we ever used,* also pronounced so at "Naval Club."

Yours sincerely, JAMAICA AERATED WATER CO., Kingston, Jamaica.

This is to certify that I have seen the original of above letter, and also many more equally complimentary.

W. B. KELLER, Editor "National Bottlers' Gazette."

FIGURE 53. Bull Ginger Ale extract was just one of the many brands available to private bottlers; the use of the Bull label was optional with the bottler.

FORMS OF ADVERTISING

The advent of the crown cap opened a new area for the labeling of the bottled product. Of all the types of seals used, the crown cap was best adapted for advertisement of the bottling firm as well as for identifying the bottled contents. Since the cap was reasonably inexpensive and used only once, the same bottle could be used for another flavor with only a change in the cap to identify the new product. Some advertising had been done previously by marking the porcelain caps of the swing-type stopper (Fig. 54). However, these closures usually listed only the bottler and seldom identified the contents.

As exemplified by the many illustrations in this book, dozens of advertising gimmicks and devices were used to better acquaint the public with the bottled products. The early use of advertising trays was a common practice of the Coca-Cola Company, especially at soda fountains in hotels, restaurants, and drug stores. The advertising budget of this firm was $100,000 as early as 1901. Although these artifacts of the better-known proprietary lines (Fig. 55) receive the greatest attention by collectors today, many of the small local firms used similar advertising schemes. Perhaps no bottling company used the advertising tray to the same extent as did the Coca-Cola firm which produced dozens of styles between the time of its origin and 1950.

The trays of the Gravel Springs Bottling Company are of equal quality (Fig. 56) but of much greater rarity because of the extremely limited production; this is true for most of the smaller bottling firms. Gravel Springs is known to have put out at least three different styles of trays. A large oval tray, undoubtedly used for serving by commercial concerns, carries the figure of a woman giving water to a child from a natural spring. This figurehead was used on all Gravel Springs advertising from around the turn of the century until sometime in the 1950s when the firm discontinued all products except mineral water. A small round tray, approximately five inches in diameter, was designed for use as a change tray and carried this same trademark. A third tin plate (Vienna Art Plate patented in 1905) was issued with advertising of the Gravel Springs Company on the bottom. This type of plate was issued in mass by the manufacturer with varying advertisements being lithographed on the back rather than on the front of the plate.

The use of posters (Fig. 57), cards and small advertising booklets, extremely popular modes of advertisement from 1880 to 1900, was employed by a number of firms. The Charles E. Hires Root Beer Company exceeded most, however, in its lavish use of this medium. Figure 58 illustrates only a few of the many advertising cards and booklets used by this and other firms. The Coca-Cola Bottling Company also advertised in this fashion but on a more restricted basis. The Gravel Springs Bottling Company used a similar form in advertising their Nix-E-Flyer during the late 1800s. It was largely due to the development of inexpensive methods of printing in color that brought about this brief but interesting stage in the advertising field. Wittemann Brothers was a firm that specialized in printing show cards, advertising cards, labels and other advertising materials (Fig. 59).

FIGURE 54. Porcelain caps, such as the swing stopper used by Hydrox, were occasionally marked with the name of the bottling firm.

During the 1880s and 1890s a proliferation in the manufacture of corkscrews with their appropriate identifying advertisements occurred. The majority of these devices were oriented toward the wine and liquor trade and not specifically for soda pop. However, the development of numerous bottle openers soon followed the advent of the crown stopper. The advertising market was opened anew and many companies found a ready market for the crown bottle opener; the Vaughan Novelty Mfg. Co. of Chicago offered "more Novelty Bottle Openers than any firm in the world" (Fig. 60).

The first patents on corrugated cardboard were issued in 1871. Initially corrugated paper was used, for the most part, for interior packing in wooden boxes. It was not until the late 1880s that the first corrugated boxes came into use. These boxes were limited primarily to local deliveries; the railroads frowned on their use as shipping containers because they could not be efficiently stacked and were easily damaged. Although used somewhat more commonly today in large shipments of soda pop, this type of container is still probably secondary to the wooden crate.

The primary container for local as well as long-distance shipping of the filled soda bottles was the wooden box or

FIGURE 55. Advertising by means of the serving tray has been used by various firms since the late 19th century. Although perfected and utilized by Coca-Cola to the ultimate, many other brands also used this medium.

FIGURE 56. The Gravel Springs Company of Jacksonville, Illinois, utilized a wide variety of advertising trays including a large oval serving tray (top left), a small change tray (top right), and a round serving tray (bottom, front and back views). The latter item was mass-produced with various firms having their advertising printed on the back.

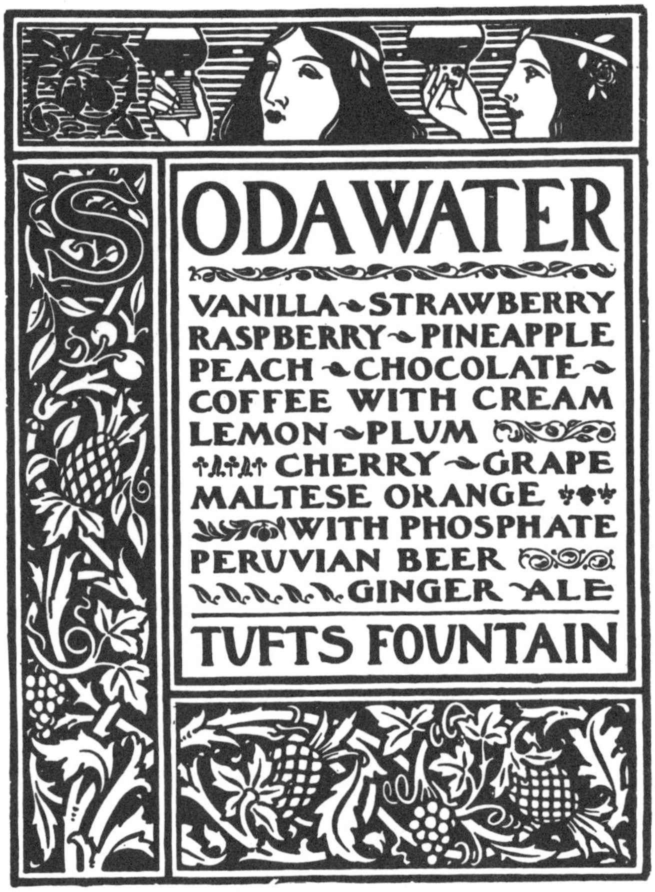

FIGURE 57. Posters, such as this one of the 1890s, were often used around the soda fountain to list flavors and to advertise various proprietary brands.

crate. Varying styles and sizes of shipping cases were available, holding from one to six dozen bottles per case (Fig. 61). However, a two-dozen-bottle container became the standard size early in the history of the business. Larger containers holding from four to six dozen were manufactured primarily for shipping large quantities of bottles for subsequent redistribution. It should be noted here that the durability of the heavy soda water bottle allowed it to be packed in a rather casual manner. The Illinois Glass Company mentions in its 50th anniversary report that soda water bottles manufactured by the firm were packed and distributed in jute sacks. Because of the thickness and compact form of these bottles, breakage was rare.

Perhaps the key—and unique—feature of these early shipping cases involved the necessity of placing the bottles in them with the stopper end downward (Fig. 62). This was necessary for several reasons. It was essential to retain moisture against the sealed cork to prevent shrinkage and the resulting leakage of gas. Although the bottoms of many of the earlier bottles varied from flat to semiround or pointed (Fig. 10), the neck and shoulder regions of the bottles were all approximately the same and allowed a rather mixed assortment to be stacked or boxed in standard containers.

Subsequent developments in the bottle seal further necessitated this type of packing. The Hutchinson stopper, for example, which protruded from the mouth of the bottle approximately one-half inch, could have been opened easily if something knocked against it. To prevent its accidental opening and subsequent loss of carbonation, it too was packed upside down with the protruding closure extending into the space in the bottom of the crate. It should also be noted again at this point that since virtually all bottles manufactured prior to 1920 had the initials or some identifying mark of the bottler embossed on the bottom, the owner could be easily determined by a quick scanning of the upended bottles. It was not until the development of the proprietary lines in the early 1920s that a particular bottle could be easily recognized when mixed with those of competitive firms. One of the rather discouraging aspects of collecting many of the soda pop bottles is the relative uniformity in style and appearance of virtually all bottles manufactured prior to 1920.

Advertising the Temperance Drink

As an apparent rebellion against the alcoholic beverage, which was perhaps the major competitor of the soda bottler, the advertisers of a number of soft drinks (especially Hire's Root Beer) emphasized the fact that their beverages were "temperance" drinks. This was done in an effort to attract those individuals of the buying public who objected to the consumption of alcoholic beverages. However, the name "Root Beer" still struck fear in the hearts of many; as only one of numerous recorded examples, the grandmother of one of the authors specifically forbade her children to consume root beer on the premise that it contained a certain amount of alcohol. The word "beer" was the key to this general attitude. Another beverage, "Malt and Hop," produced by the Gravel Springs Company of Jacksonville was advertised as looking, tasting, and smelling like beer but being nonalcoholic. Advertised as a pleasant substitute for beer, it was considered a bracing tonic which under constant use would strengthen the system.

It must be noted, however, that some of the early carbonated beverages sold largely as soft drinks did contain a small proportion of alcohol. Apparently the alcoholic content varied considerably, depending on the discretion of the local bottler and the amount of time that passed between bottling and consumption. Additionally, many of the early soda bottlers also bottled alcoholic beverages, and even this association often distressed temperance advocates. Letterheads of two southern Illinois companies (Fig. 63) typify the efforts of many firms to emphasize their products as being temperance drinks.

Transporting the Product

Some mode of transporting beverages from the bottling plant to consumers, especially after the soft drink "fad" caught on in earnest, became a basic necessity for the bottler. The delivery vehicle also served a secondary function, that of advertising the product and bottler (Figures 64 and 65). However, if the product could not be delivered to the consumer, then all the advertising in the world would be to no avail. The growth of the industry essentially paralleled the development of a suitable vehicular conveyance and the increase in the potential number of customers available. The earliest delivery wagons, pulled by draft teams, were able to supply consumers within a relatively limited region. It should be noted that, during these

FIGURE 58. Advertising cards, especially popular from the 1880s to about 1900, were widely used by some firms. Root beer extracts were extensively advertised in this fashion and the cards were frequently handed out as give-aways at grocery stores selling the extracts.

FIGURE 59. The paper label was one of the earliest means of advertisting the bottler and identifying the bottled product. Specially designed labels could be purchased from numerous firms such as Wittemann Brothers (*N.B.G.* 1886).

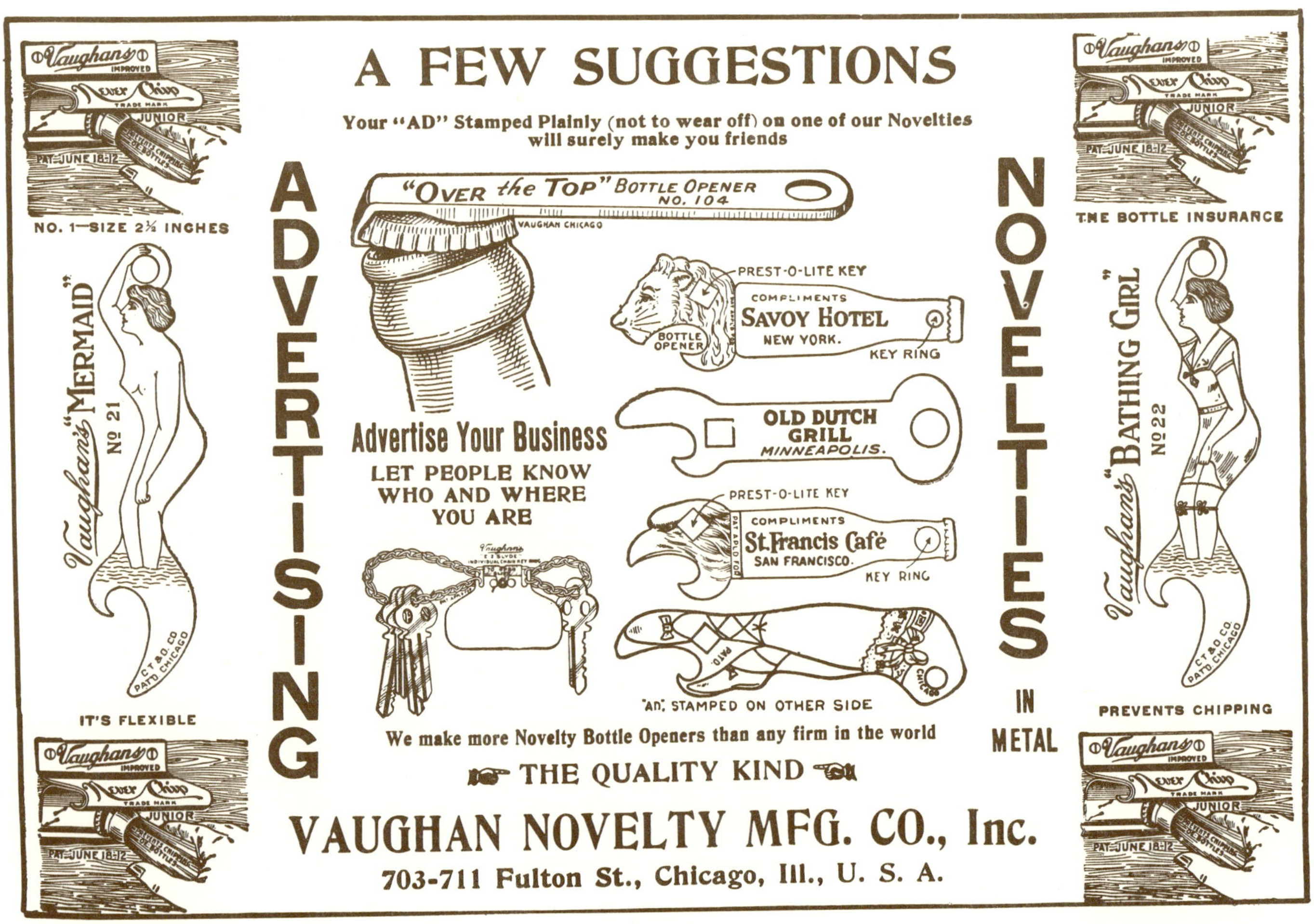

FIGURE 60. With the development of the crown closure, a proliferation of bottle openers soon followed. These provided another means of advertising the bottler and his product (*N.B.G.* 1920).

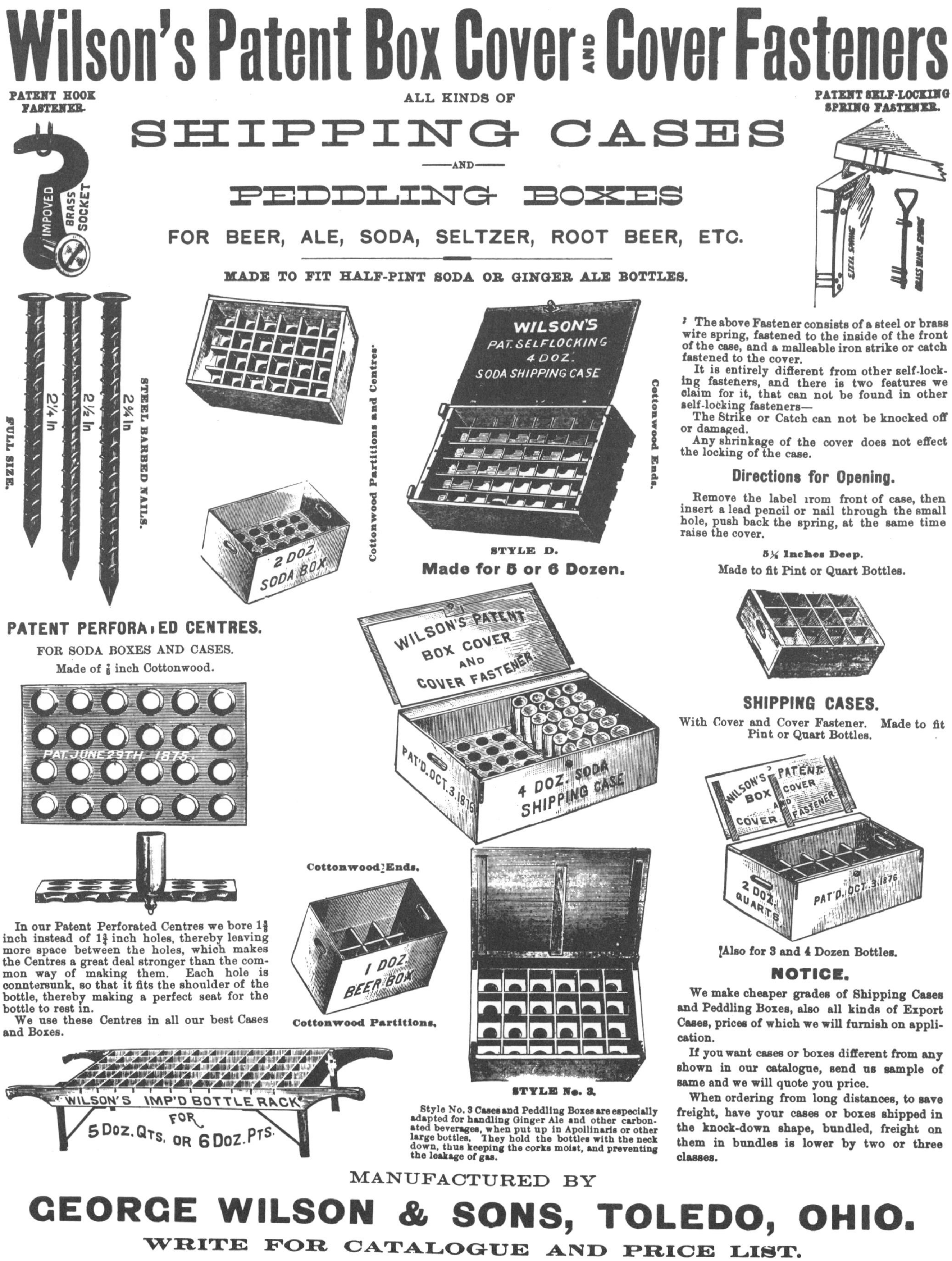

FIGURE 61. Sturdy packing and shipping cases were important pieces of equipment in the operation of a bottling business, and many box companies produced cases and accessories for every need (*N.B.G.* 1885).

FIGURE 62. Cases served as a means of advertising the bottler as well as to transport and protect his product.

earliest days, the soft drink business was primarily a summer occupation; most plants locked their doors shortly after Labor Day and did not reopen them until the warm days of spring. The season was lengthened somewhat as expanded market areas were opened. Nevertheless, the primary limitation of the soft drink business during these early days was the delivery of a liquid beverage that was susceptible to freezing. A bottler simply did not load a wagon with such a fragile cargo when the temperature was substantially below freezing; even when his destinations were only a short distance away, losses from freezing would quickly consume the profits.

Suggestions for overcoming this drawback were made by various individuals. The conical-shaped bottle produced by E. J. Salisbury of Mt. Clemens, Michigan, was an attempt at developing a container that would allow enough space for expansion resulting from freezing (Fig. 66). However, even in these modern days, there was a period during the winter of 1970-71 when the distributors in Springfield, Illinois, literally exhausted their supplies of

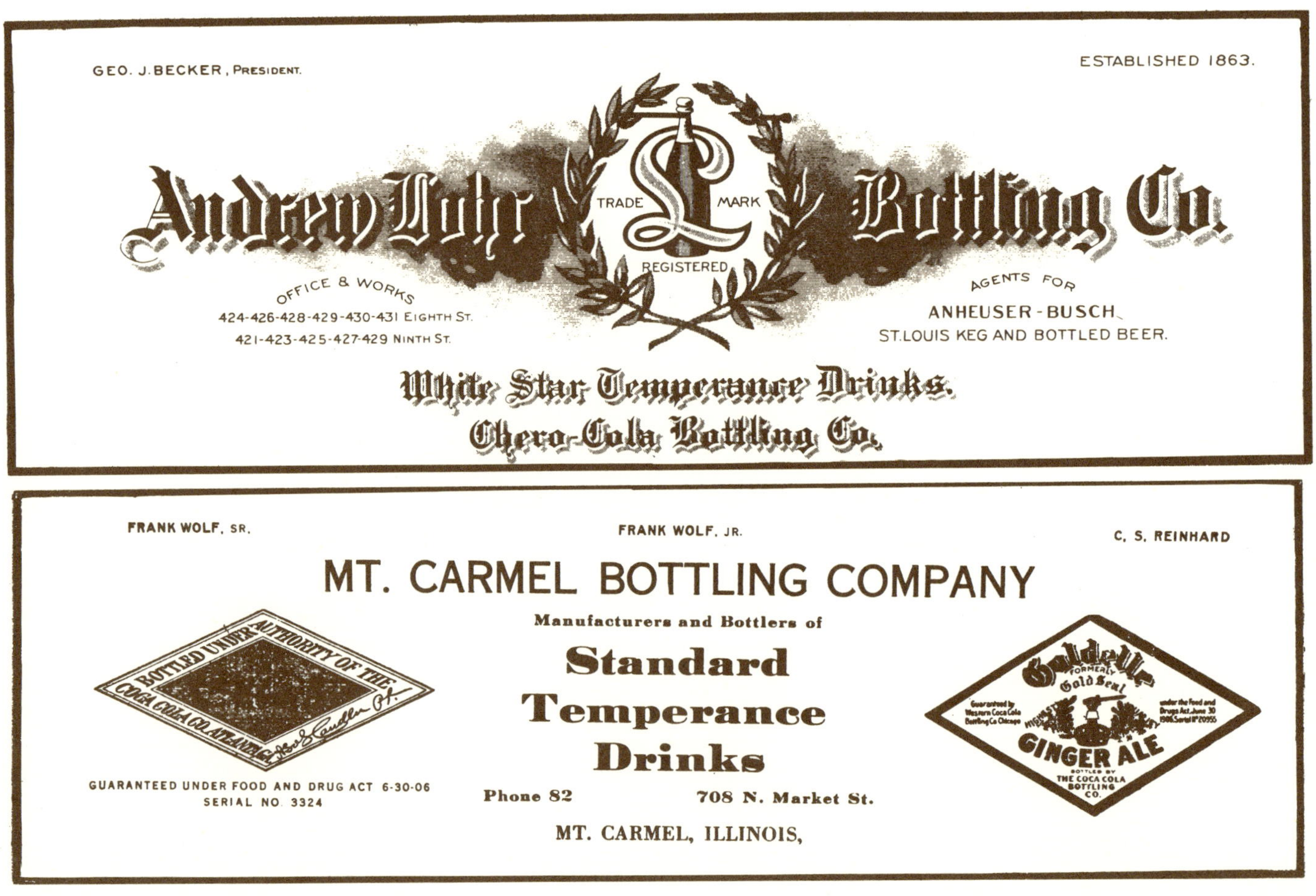

FIGURE 63. Many of the early bottling firms, as evidenced by their letterhead, stressed the temperance aspect of their soft-drink products.

Figure 64. This photograph (ca. 1920) shows delivery vehicles of the Coca-Cola Botting Company, Mt. Vernon, Illinois. Trucks were a necessary mode of transportation of the small-town bottler for distributing to a regional trade (Photo, courtesy William R. Wolf).

several brands of soft drinks because of extreme cold weather conditions. The only major distributor in downstate Illinois for Coca-Cola products is at DuQuoin, some 125 miles to the south; during periods of extremely cold weather, no beverages were transported from the bottling plant to the local distributor. Thus, after 100 years of bottling and delivery experience, extremes in weather conditions are still a factor.

The replacement of the horse-drawn wagon by the motor truck significantly enhanced the ability of the bottler to sell his product over a wider geographic area. Most bottlers had incorporated truck transportation into their business operations by the 1915-to-1920 period, and by so doing they could adequately service retail outlets within a 20- to 30-mile radius of their bottling plant. It was essentially during this period, one characterized now by the development of improved transporation, that a substantial increase in sales of the product was made through the corner groceries, saloons, candy shops and other neighborhood outlets. Prior to this time, from about 1880 to 1920, the greater majority of products were delivered directly to the consumer and were not sold through an intermediate retailer. Simultaneous with the development of improved transportation facilities was the corresponding improvement of the road system which did much to facilitate larger and more distant deliveries.

Chero-Cola Bottling Companies Own 234 White Trucks

SIX years ago, four White Trucks were operated by Chero-Cola Bottling Companies. Today, 234 White Trucks are owned by these companies. This rapid, steady growth is evidence of the White's ability to render fast, sure and low cost service to the bottler. White Trucks are everywhere known for their high earning power. Comparative records show that they do *the most work for the least money.*

Representative Expressions from Chero-Cola Companies:

"We would not consider buying anything but White, as we consider that make better than any we have ever used."
CHERO-COLA BOTTLING CO.,
Fitzgerald, Ga.

"No bottler will go wrong in purchasing White Trucks. They are always on the job and you can depend on them to deliver the goods."
CHERO-COLA BOTTLING CO.,
Augusta, Ga.

"To the bottler, delivery of goods is a most ital question. White Trucks solved this question for us by proving that our delivery costs could be reduced to a minimum."
CHERO-COLA BOTTLING CO.,
Chattanooga, Tenn.

"White Trucks are more economical and longer lived than trucks of any other make."
CHERO-COLA BOTTLING CO.,
Charlotte, N. C.

"We have used nine other makes of trucks. At present we are using nothing but Whites and know that two Whites will do as much work as three trucks of any other make."
CHERO-COLA BOTTLING CO.,
Madison, Ga.

"We have been using White Trucks for two years and have saved the difference in the first cost of Whites and cheaper trucks over and over again each year in an absence of repair bills, lower operating cost and better service."
CHERO-COLA BOTTLING CO.,
Spartanburg, S. C.

THE WHITE COMPANY, *Cleveland*

White Trucks

FIGURE 65. Competition between various truck manufacturing companies (White trucks shown above) to secure the bottler's business for their firms was often vigorous (*N.B.G.* 1920).

THE ONLY BOTTLE

SUITABLE

FOR WINTER SHIPMENTS.

SALISBURY'S PATENT.

WHEN CONTENTS FREEZE THIS BOTTLE WILL NOT BURST IF FILLED AS DIRECTED.

No More Tedious Delays in Winter.

YOU CAN NOW SHIP BOTTLED GOODS WHEN YOU PLEASE.

Salisbury's Shipping Bottle Defies Jack Frost

WHEN STOOD ON ITS HEAD.

Anyone can understand the principle involved by looking at the cut. It is a conical bottle, filled to a given height, and then inverted before exposure to cold. The rings on the neck indicate to what height the bottle is to be filled with different liquids. The lower ring marks the amount of highly charged liquids (those bottled at over 10–15 pounds pressure), which may be introduced, while the upper one is intended for beer, ale, porter and still liquids, such as cider, vinegar and in fact any liquid containing water and which is liable to freeze. When the contents become solid, the inverted cone of ice formed has a chance to expand in an upward direction without breaking the bottle, owing to the shape given to it.

Allowance has been made for expansion of the liquid in freezing, and also for gas pressure. When the contents are thawed out a simple shaking up restores them to their original condition.

This Bottle has been subjected to Crucial Tests during the severe Winter of 1884-5, with all kinds of goods—beer, ale, porter, mineral waters and soft drinks. These drinks lose none of their properties, and are as good as ever when thawed out and shaken up.

The Bottle costs no more than any other style of same size. It has a symmetrical and attractive shape, and makes quite an ornamental bottle for the bar.

Owing to its being patented, the user is protected from the usual loss by stealage where bottles are loaned out. This fact alone gives it an important advantage over ordinary styles.

As an advertising feature it is decidedly valuable. The fact of being able to ship your goods with safety when the thermometer is below freezing point, while your competitor's hands are tied waiting for a change of weather, gives you an immense advantage. This advantage accrues in turn to **your** customers. For Summer use it can be packed very nicely in the ordinary way.

STATE AND COUNTY RIGHTS FOR SALE.

E. J. SALISBURY, Patentee,

Mt. Clemens, Mich.

FIGURE 66. The Salisbury patent bottle, while providing a solution to winter freezing problems, apparently received only nominal interest by the trade (*N.B.G.* 1886).

Chapter IX
OPERATIONS OF THE PRIVATE BOTTLER

The profitable business potential offered by a new and rapidly expanding soda pop trade attracted a diverse but enthusiastic group of individuals with varied backgrounds. However, many of the earliest successful bottlers initially began the venture as druggists, serving soft drinks at their fountains. The successful development and manufacture of more preferred flavors, added improvements in the basic machinery and procedures used to prepare soft drinks, and eventually better bottles and bottling techniques allowed these "amateurs" to change to this entirely new sideline. For example, James Vernor of Detroit, Michigan, began in the 1880s as a druggist, giving up this trade when his beverage, "Vernor's Ginger Ale," became one of the most popular drinks of that region. James W. Tufts of Somerville, Massachusetts, one of the leaders in the manufacture of soda fountain apparatus, began his endeavors in 1863 as a result of a need for an improved soda fountain in his drugstore.

The available market for new and better equipment and supplies greatly exceeded the profit potential of the corner drugstore. Other individuals made equally historic and significant developments, one of the most notable being the Hutchinson firm of Chicago. In these earliest years, it is said that this company hauled all of its water for carbonating in barrels from Lake Michigan. As the town grew and business prospered, T. O. Dunn joined as a partner in the firm in 1858. This partnership continued until 1864 when Mr. Dunn sold his interest to two of the Hutchinson sons. The firm thereafter was known as W. H. Hutchinson and Son. Upon the death of the elder Hutchinson in 1879, one of his younger sons, Charles G. Hutchinson, entered the business and immediately patented the famous Hutchinson Stopper. The demand for this stopper was so great that, in 1882, the soda and mineral water business was sold to Hayes Brothers of Chicago so Hutchinson could devote full time to the manufacture of the now widely used stopper and other bottler accessories. It is of some historical interest to note that immediately after the Chicago fire in 1871, the post office, which had been totally destroyed, borrowed wooden shipping cases from the Hutchinson firm for use in the sorting and distribution (pigeonholing) of the mails. By adding a line of extracts and general bottling supplies to the manufacture of the now-famous Hutchinson stopper, the business continued to grow during the 1880s. Around 1887 the fountain portion of the business formed the basis of a new firm called the Chicago Fountain Soda Water Company of which Charles G. Hutchinson became president.

As mentioned in the section on closures, the development of an accepted stopper was not a simple matter in those early days. Patent fights were the norm and only through extended legal battles did the Hutchinson firm establish its patent priority on its stopper. With the expiration of the patent rights on the stopper in April, 1896, the firm was well-enough established to maintain itself as a supplier of all necessary materials to the soft drink bottler.

Riley (1957:162-163) presents a detailed description of a typical small-town bottling plant at the turn of the century, the Wagner Bottling Works of Venice (now Granite City), Illinois. This firm, organized in 1901 by J. Fred Wagner, was typical of small plants of its day. Riley offers the following observations:

> The flavoring syrups were prepared in stone crocks, mixed by wooden stirring paddles. When ready, the crock's outlet was connected by a flexible hose, or pipe arrangement, with the syrup valve on the filler. Liquid carbon dioxide was used; [This was a rather modern innovation and not typical of the small firms who had been bottling for any number of years.] The filler was a foot-power Crown Cork and Seal machine, the bottles and crowns being hand-fed. [This too was a "modern" device.] Bottles were washed in wooden tubs, by hand; the first automatic power washer was not installed until 1910. Water treatment was simple; for the first few years a stone filter was found adequate for clarification. In 1910 a water still was installed, and the distilled water was used for bottling and for other plant operations. Flavors were the usual white, black, and red line — meaning lemon, sarsaparilla, and strawberry. Delivery was made by a horse-drawn wagon, to a community of about 15,000, with sales amounting to 12,000 cases the first year. It was not until 1903 that the second team and another wagon were added to the delivery equipment. The first auto truck was used in 1915. And, about 1910, the first juice drink — orange — was added to the line, as well as a nationally advertised beverage known as "Iron Brew." Principal outlets were the community's pool rooms, restaurants, coffee houses, fishing resorts and picnics. It was a summertime business.

An interesting brochure advertising the potential profit that might be realized from a small plant was circulated by the Cairo Food Products Company of Cairo, Illinois (Fig. 67). This ad is particularly noteworthy because it not only provides a price list of equipment and supplies necessary for such a small early 20th century plant but also points out a new flavored drink, Mint Cola, as "The Best Money Maker." Such advertisements outlining a method for establishing a profitable business with relatively little investment undoubtedly encouraged many to enter the soft drink trade. The fact that August is not considered as part of plant operation time because it is "watermelon month" is especially intriguing!

The Moxie Plant of New York City, a comparatively large-capacity business operating during the same period, is briefly described as follows (Riley, 1957:162-163):

Mint Cola

THE BEST MONEY MAKER

Earning Power of a Small Bottling Plant

Fully Equipped Under Efficient Management.

Figures From Actual Experience

300 Cases output per day, (2 Doz. each) @ 70c Val.		$210.00
Cost of Syrup $75.00, Crowns $16.50, Gas $6.00	$97.50	
Cost of Water, Breakage and Loss	8.00	
Cost of Labor, 4 Men, $3, $2.50 $2, $2	9.50	
Overhead	12.00	
Total Cost 300 Cases		$127.00
Daily Profit		$ 83.00

200 Cases Output per day (2 Doz. Each) @ 70c Val.		$140.00
Cost of Syrup $50.00, Crowns $11.00, Gas **$4.00**	$65.00	
Cost of Water, Breakage and Loss	6.00	
Cost of Labor, 3 Men, $3, $2, $1.50	6.50	
Overhead	10.00	
Total Cost 200 Cases		$93.50
Daily Profit		$46.50

100 Cases out put per day (2 Doz. Each) @ 70c Val.		$ 70.00
Cost of Syrup $25.00, Crowns $5.50, Gas **$2**	$32.50	
Cost of Water Breakage and Loss	3.50	
Cost of Labor, 2 Men, $300, $2.00	5.00	
Overhead	8.00	
Total Cost 100 Cases		**$49.00**
Daily Profit		**$21.00**

Now suppose the plant is run only six months out of a year, or, say during April, May, June, July, September and October, or, 150 days. (August is skipped because it is watermelon month.)

Your Profit would figure as follows:

300 Cs. Daily Output for 150 days @ daily profit	$83.00	$12,450.00
200 " " " " " " " " "	46.50	6.975.00
100 " " " " " " " " "	21.00	3,150.00

In addition to the above you can figure on doing at least half as much more during the rest of the year, as a good Cola Drink sells the year round, but we have left that out for good measure.

Cost to Equip a Little Plant and Start Right as Follows:

1 Foot-power Crowning Machine	$ 240.00
1 Carbonator	350.00
1 Soaker	550.00
1 Electric Motor	175.00
Plumbing and Setting up	100.00
250 Gross Bottles @ $4.50	1125.00
900 City Crates @ 40c each	360.00
100 Shipping Cases @ $2.50	250.00
Syrup Hose, etc.	150.00
Percolator	60.00
Horse and Wagon	350.00
Total	**$3710.00**

Correspondence Solicited

CAIRO FOOD PRODUCTS CO.

CAIRO ILLINOIS

FIGURE 67. Many small firms attempted to break into the market with various proprietary brands. The statistics put forth by the Cairo Food Products Company for their "mint cola" brand were enticing to the individual who wanted to start his own business. Note the recommendation to discount August as a sales month because of competition with the watermelon.

The plant was devoted entirely to the bottling of one of the early proprietary drinks. All the water was filtered, then distilled and cooled prior to carbonating; again, liquid CO_2 was used in the plant. Although the plant's volume of production is unknown, the filling line was apparently large, and consisted of foot-powered bottling machines which utilized crown closures. The crowns were hand-fed into the machine. Flavoring syrup was piped from a mixing vat to each of the bottling machines where automatic syrup gauges allotted the correct amount to each bottle. All bottles were washed with filtered water. Following the soaking operation the brushing-rinsing operation was done on a machine having 16 operating heads. One of the early types of gravity-operated roller conveyors was used to move the cases through the plant and to and from the filling line.

A historical study of the beginning, development and the eventual fate of each of the hundreds of private bottling firms that flourished in Illinois cities and towns from the 1870s to around 1930 would make a fascinating story in itself. All shared similar problems relative to equipment, supplies, production and the distribution of their products, but only a relatively few remained in business for longer than a decade. The reasons for success or failure were many and varied. A good number of these early bottling firms started as a one-family operation; therefore, whether the operation continued in business after the founder died or retired depended on his children's interest in the company. A favorable location to sources of supplies, labor and transportation were of vital importance as was the growth of the city or area in which a firm first started in business. Installation of new and modern equipment, plus greater production of more diversified soft drinks, often spelled the difference between success and failure. Extensive and varied advertising came to play an important role as well. The following brief descriptions of the operation of two successful private Illinois bottling firms typify many of those which flourished during the early decades of the 20th century.

The Gravel Springs Company, Jacksonville

The Gravel Springs Company of Jacksonville, a small west-central Illinois city, may have been the most successful bottler of mineral water in Illinois. Started in 1873, the firm is still operating today although sales are restricted currently to bottled water from the spring; the subsidiary soda pop lines have been discontinued. Located in Morgan County near the site of a natural spring with a flow rate of 3,000 gallons per hour, the firm was advertised as having one of the purest natural waters available. In the beginning water was transported from the spring by horse-drawn wagons into Jacksonville where it was placed into appropriate containers and shipped to varying markets. Business boomed, however, and to facilitate output, a pipeline was laid connecting the springs directly with the bottling plant in downtown Jacksonville, approximately two miles away. The firm did such a volume of business that four railroad systems had spurs to the plant and whole tank cars were filled for distribution to the larger cities such as St. Louis and Chicago.

In addition to the bottled natural mineral waters, sales included several "doctored" waters in which additional mineral salts were added to obtain the prescribed (and well-advertised) medicinal effects. A full line of soda pop was also carried, although ginger ale was emphasized during the earliest years of the firm's operation. The natural spring water was available in tank-car lots, ten-gallon cans, and five-, two- and one-gallon jars. Most of the other beverages were bottled only in quarts, pints, and splits, the latter two sizes being available in cases of 100 and quarts in cases of 50. Other relics of its past include a small drinking glass (Fig. 68) and a crockery dispenser with spigot which was apparently used either in soda fountains or in business offices for dispensing the "refreshing" water.

The Gravel Springs Company apparently was a most successful business even though restricting itself largely to Illinois sales. Its economic impact was sufficiently great in the Chicago area that the firm of Hinckley and Schmitt, a Chicago bottler and distributor of mineral waters (Fig. 69), was reputed to have paid the Gravel Springs concern $25,000 in return for not distributing its products in the Chicago region. Apparently the Gravel Springs firm accepted this offer sometime between 1910 and 1915 and, by so doing, restricted its sales after that period to the downstate region.

Additional descriptions of this company's products are presented in the section dealing with the medicinal qualities of mineral and soda waters (Chapter I) and in the section under advertising where the trays and other effects of this firm are discussed (Chapter VIII).

The Mt. Carmel Bottling Company, Mt. Carmel

One of the oldest firms in Illinois to operate continuously under the control of a single family is the Coca-Cola Bottling Company of Mt. Carmel, Illinois. Organized in 1887 by Frank J. Wolf, the business is currently run by William R. Wolf, grandson of the founder. The initial bottling business was known as the Mt. Carmel Ice and Bottling Company when the operation combined the sale of ice cut from the Wabash River with the summer business of selling soda pop. In the earliest days it was necessary to sell other products, for example, ice and vinegar, along with soft drinks. In subsequent years, beer was added to and maintained in the company's line.

In 1905 Mr. Wolf became aware of a relatively new product called Coca-Cola, and he made arrangements with the Coca-Cola Bottling Company for a franchise. Although the sale of Coca-Cola was slow during the first three or four years, the drink soon became popular and the franchise was granted to cover the Mt. Vernon and Harrisburg territory in addition to the original local sales area.

It is interesting to note that these early franchise territories were largely defined by the railroad lines within the areas they served. In the late 19th and early 20th centuries, the standard method of distributing the beverages was by railroad. For example, cases were loaded on a railroad car at Mt. Carmel in the morning. Mr. Wolf would then travel with the drinks to the towns along the line; each day he would stop at one or more towns, unload the soft drinks from the train, rent a wagon locally for the purpose of delivering the full bottles and picking up the empties. In the evening he would return

FIGURE 68. Drinking glasses such as this one put out by the Gravel Springs Company of Jacksonville, Illinois, also served as a medium of advertising.

by train to Mt. Carmel, ready to go the next day to a different area. It was only through the relatively rapid transportation provided by the railroad that bottlers were capable of serving more than a single locality within their franchised areas.

Correspondence on file with the Wolf Company indicates that some of its competition from surrounding towns had, on more than one occasion, complained to the Western Coca-Cola Company of Chicago which issued the franchises. They argued that the Mt. Carmel plant should not be given certain portions of their franchised areas because of the difficulty in servicing them since adequate rail facilities were not available

In 1913 the Mt. Carmel Bottling Company purchased the bottling plant formerly owned by Henry Dietz, Jr., of Carmi. Of special interest is a complete list of apparatus which exchanged hands. The Dietz plant equipment and supplies, as listed in the purchase agreement, consisted of the following:

> One (1) Carbonator, one (1) Crown Soda Bottling Machine, one (1) Labeling Machine, one (1) Gillett Shot Bottle Washing Machine, one (1) Washing Tub, two (2) Tanks (one large and one small), one (1) Seltzer Bottling Machine, all bottles and boxes, Syphons formerly owned by Henry Dietz, Jr., syrups, extracts, crowns and labels, not including Motor and Shafting on account of a dispute between John V. Heitzman and Henry Dietz, Jr., three (3) Filters, one Small Ice Chest, two (2) 50 lb. Cylinders Gas (one partly used), also Water Heating Apparatus with tank, also plumbing used for the Bottling outfit, syrup jar and stand, one (1) one-half gallon Copper Measure, one (1) gallon Copper Measure, one (1) Granite Syrup Mixer, one (1) Drum Stand and Gauge, one (1) Clock, two 12 Gal. Syrup Jars.

This list provides a rather good insight into the types of equipment used by the small plant at this time. It was somewhat more advanced than many of the older firms in that it used crown bottles, thus suggesting that this plant had not been in operation long before the turn of the century. Figure 70 shows some of the bottle types used by the Mt. Carmel Bottling Company during its many years of operation. Included is an early Coca-Cola bottle in which the script lettering and design was done by Mr. Frank Wolf himself. It was not until the advent of the characteristic Coca-Cola bottle, first patented in 1915, that uniformity of containers was enforced.

The firm continued to carry beer after the repeal of prohibition until about 1939 although the Coca-Cola Company felt that beer was harmful to the sales of their soft drink. However, up to that time, both beer and Coke had been carried, on occasions, on the same delivery truck.

FIGURE 69. Hinckley and Schmitt of Chicago, Illinois, were major distributors of mineral water in the late 19th and early 20th centuries.

FIGURE 70. This array of bottles represents many years of bottling by a single family (Wolf of Mt. Carmel, Illinois). Although most are proprietary styles of the 1920s and 1930s, the bottle at left is an early Coca-Cola bottle designed by the founder of the firm prior to the development of the now-famous Coke bottle.

Hazards of the Trade

All businesses are not without some dangers and frustrations in one form or another, and the private soft-drink bottler at the turn of the century was no exception. There were numerous reported incidences of accidents, thefts, fires and even personal assaults, all in the line of duty; these accounts point up the fact that the early bottling industry boom was not all pleasure and profit!

One of the monthly columns in the *National Bottlers' Gazette* was labeled "Casualty Chronicle." A perusal of these listings provides some interesting and often humorous reading, although their purpose was to relate the potential dangers within the bottling profession. The following examples were extracted from various 1896 issues of the *National Bottlers' Gazette*:

> Jere Deegan, Pottsville, Pa., was seriously injured by being thrown from his wagon on the 22d ult. A rope, stretched across the street in which he was driving his loaded wagon, and unnoticed, swept off the body of the vehicle, with Mr. Deegan, who was picked up senseless in the wreck of smashed bottles. The horses ran away.
>
> Frank Rheme, Shelbyville, Ind., was severely cut on the forehead, on the 21st ult., by the bursting of a bottle in a box he was carrying into a customer's place. [Although far less common today, this danger still exists.]
>
> George Nichols, a traveling salesman for Dean, Foster & Co., glass manufacturers, Boston, is taking an involuntary whaling voyage. Last June he disappeared mysteriously, and in December his father, of the wholesale drug firm of Nichols & Harris, of New London, Conn., received a letter from the missing man, relating a curious story. It appears that upon the day in June when young Nichols disappeared, he joined a party of chance acquaintances, who boarded a tug in New Bedford harbor. The tug was chartered to tow out a whaling vessel, the James Arnold, bound on a two years' cruise in the Pacific and South Atlantic, and intended to round the Horn. When open water was reached the tug was cast off, and she steamed ashore with all but Nichols who had fallen asleep on the whaler and was not missed by his party. When Nichols awoke he found himself practically a stowaway, miles from land. He made the best of the situation, however, and cheerfully took hold of the work with the regular crew. So there is where the natty drummer is now, cutting blubber, and working, doubtless, harder than he ever worked in his life. Nichols is 30 years old, and widely known throughout the east.

The determination of a bottler and a reflection on the times are revealed by the following incident:

> Springfield, Illinois
>
> J. Rule of Rule Brothers Bottling Works was making his rounds in the north part of the city when he was hailed by a man who said he wanted to speak to him. Rule got off the wagon and approached the sidewalk where the man stood; as he did so, the stranger drew a knife and ordered Rule to throw up his hands, let him have his money or he would cut his heart out. Rule's hands went up immediately (for business). The first blow landed his assailant on his back several feet away. He sprang at Rule a second time cutting a slit two and one half inches long in Rule's shirt right over his heart. Rule downed his assailant again and proceeded to give him a drubbing to be remembered. In the mele Rule's right leg was stabbed just above the knee, reaching the bone and breaking the blade off the handle. After disentangling himself, Rule pulled the knife blade out of his leg, remained on duty and was all right in a day or two.

In summary, day-to-day operational hazards so commonplace in a turn of the century bottling plant might best be exemplified by the statement printed in the July 1901 issue of the *National Bottlers' Gazette*:

> Walter Clutter, an employee at the Springfield [Illinois] Mineral Spring Works, North 6th Street, was injured when a bottle of soda exploded, blowing his left arm full of glass.

Lastly, the following comment that appeared in the *National Bottlers' Gazette*, June 1896, seems timely in light of some of today's social problems:

> The Illinois Supreme Court on the 12 ultimate sustained the action of the druggist who refused to sell soda water to a Negro. The court held that the selling of soda water does not constitute the furnishing of lodging, amusement or meals.

Chapter X
ILLINOIS BOTTLERS AND BOTTLES

The number of private, local bottling companies that flourished at one time in Illinois during the period from about 1870 to 1930 seems almost endless. All that remains today of many cities that supported industry and numerous small businesses — including one or more bottlers — are a few homes and a country grocery store. The possibility of locating and recording all of the early Illinois bottling firms, either through published accounts, reminiscences of older members of a community, or by recovery of bottles used during a company's operation, is indeed remote. Toward this end, however, information supplied by bottle collectors and other individuals interested in the subject have added immeasurably to our own records of early bottlers. In Illinois, from about 1870 through 1930, our records show that 116 cities had at least one bottling plant, 78 had from two to five, and 28 cities had six or more different bottlers. Chicago, alone, had over 200 bottlers and/or bottling plants. As time passes, these figures will surely increase with the discovery of long-forgotten records — and the recovery of buried bottles.

Some bottling firms survived for over half a century from the time of their initial operation while others were extremely short-lived. Each company is a fascinating study in itself, but to research the complete history of even a few would involve a tremendous amount of time. Records of firms, especially those in operation prior to and at about the turn of the century, are often incomplete or nonexistent. The style or type of bottle often provides the only clue as to the time period a particular firm was in operation. For example, the Hutchinson "blob" was in use from about 1885 to 1910 and the applied crown from about 1905 to 1920 until it was gradually replaced with the crown soda bottle produced automatically in one process. Some companies are known to have continued using the Hutchinson bottle even after the crown had come into general use. Bottles with applied crowns were also used until the stock was finally depleted through loss and breakage.

Numerous companies went through various periods of change as the process of manufacturing and bottling soda pop was perfected. Development of more efficient machinery, better flavorings, rapid processing, and availability of proportionately cheaper bottles made it possible for many firms to expand their business. Some of these changes, necessary in order for a bottler to continue a profitable business, were factors that often affected company ownership or management. An individual would go out of business, for one reason or another, and sell his entire plant, equipment and stock. Often one bottler would purchase another's business as a means of expanding his own and eliminating local competition. Occasionally, owners of separate companies would combine their assets and operate the "new" firm under joint management. These changes are often reflected in the names embossed on bottles: John Lauterbach/Lauterbach and Reisch; C. J. Peterson/Johnson and Peterson; Marshall Bottling Co/Davis and Marshall; and M. A. Fischer/Fischer and L. Davis — all of Springfield (Ill.) — are only a few of many such examples. On the other hand, a firm might remain under the control and management of one family for several generations and, as in the case of the J. J. Flynn Company of Quincy (Ill.), all bottles from Hutchinson to the proprietaries of the 1920s and 1930s are embossed J. J. FLYNN & CO., QUINCY, ILL.

Soda pop and mineral water bottles were manufactured strictly for a utilitarian purpose, and it cannot be said that their basic sizes, colors and shapes make them objects of great beauty. Nevertheless, for several reasons these rather unimpressive-looking bottles have appeal for numerous collectors. Embossed company, city and state names, plus occasionally an interesting trademark emblem such as an animal or ship, make them desirable. Most of the popularity of bottle collecting centers around embossed bottles, whether they are medicines, beers, whiskeys, soda pops or flavoring extracts. Soda bottles are of certain historical interest since they often reflect a bygone era and a business name which has all but been forgotten. This is especially true of the Hutchinson and pre-Hutchinson bottles although their age and the hand-applied tops are other factors which make them collectible. Although of more modern vintage, the proprietaries of the 1920s and 1930s, with their varied shapes, sizes and designs, also have considerable collector appeal. Because of these factors which have made these old, otherwise useless, bottles known, it has been possible to locate and date many of the early Illinois bottling firms that have long since passed into unrecorded history.

The following listing of Illinois bottling companies is based primarily on extant bottles that have been examined and recorded and on information obtained from the *National Bottlers' Gazette*, various city directories, manufacturing census records, and individuals knowledgeable in the history of a particular city or county. This list is by no means complete, but it does reflect a concerted effort to bring together in brief catalogue form the names, locations and inferred periods of operation of Illinois bottlers. Emphasis has been placed on the bottles; the basic styles, color, and relevant embossing are recorded. Slight variations in embossed wording and word arrangement often occur on a style of bottle used by a particular company, but these minor variants seldom warrant separate listings. Although crown proprietary bottles (ABM-P) were produced on automatic bottling machines, most are distinguished here from the earlier, simply styled and plainly embossed machine-made crown soda bottle (ABM). Some of the earliest crown soda bottles were made on semiautomatic machines, for example, those produced by Root, but all the machine-made crowns are listed as ABM. The companies are listed alphabetically by city; county names are in parentheses after each city name.

Key to Abbreviations

Source of Information

CD City Directory
MC Manufacturing Census
NBG *National Bottlers' Gazette*

Basic Styles

ABM Automatic Bottle Machine, Machine-finish Crown
ABM-P Automatic Bottle Machine, Proprietary Bottles
AC Applied Crown
HB Hutchinson Blob Top
TB True Blob Top

SOFT DRINK AND MINERAL WATER BOTTLERS OF ILLINOIS: circa 1860 - 1930

Embossed Company &/or Bottler Name. Comments: (Owner's name, date(s) of operation, ownership changes, etc.)	Source of Information (Plate Illus.)	Bottle Style	Bottle Color	Vol.	Major Bottom Marks	Trade Mark	Bottle Manuf.
ALEDO (Mercer Co.)							
Zales S. McGinnis Co.	Bottle	ABM	Aqua	$7\frac{1}{2}$oz.	McG		
ALPHA (Henry Co.)							
Alpha Bottling Works	Bottle	AC	Aqua				
ALTAMONT (Effingham Co.)							
Altamont Bottling Works	Bottle(78)	AC	Aqua		F		R G Co.
Altamont Creamery	Bottle(78)	ABM-P	Clear	$6\frac{1}{2}$oz.			
(C.E.Munday, beginner)	NBG,3/1891						
ALTON (Madison Co.)							
A. P. Ernst	Bottle	HB	Aqua				
A. Ioerger & Bro.	Bottle	TB			Iron pontil		
(C.Weisbach, soda water bottler)	1880 MC						
(Weisbach sold to Ernst)	NBG,1/1892						
Max Kuhl	Bottle(71)	HB	Aqua				IGCo.
Lampert Bottling Works	Bottle	AC	Clear	8 oz.	F & L		
Lampert Bottling Works	Bottle(78)	ABM	Green	8 oz.	F & L		
C. O. Marshall	Bottle(71)	HB	Aqua		M		
Herman Meyer	Bottle(71)	HB	Aqua				IGCo.
(Meyer sold factory to Charles Graff and Fred Nicolet)	NBG,ca.1891						
(John C. Misegades, soda & mineral water bottler)	1905 CD						
Schmidt & Knecht	Bottle(71)	HB	Aqua		S & K		IGCo.
C. Weisbach	Bottle(71)	HB	Aqua		C.W.		A & DHC
(Reference to Christopher Weisbach)	NBG,6/1891						
(Jacob Weisbach, soda water bottler)	1860 MC						
(Christ. Weisbach, bottler)	1868-69, 1874 CD						
(Weisbach Bros., soda & mineral water bottlers)	1905 CD						
A. Yackel	Bottle(71)	HB	Aqua		Y		
(Adolph Yackel, soda & mineral water bottler)	1905 CD						
ANDALUSIA (Rock Island Co.)							
Renna Wells	Fitch 1927						
ANNA (Union Co.)							
Western Saratoga Spring	Fitch 1927						
ARGO (Cook Co.)							
Argo Bev. Co.	Bottle	ABM	Green	7 oz.	Argo B.Co.	Pear	
ARLINGTON HEIGHTS (Cook Co.)							
F. W. Muller	Bottle	ABM	Aqua	6 oz.	Bottle	Initials in star	
F. W. Muller Sons	Bottle	ABM	Clear	23 oz.		M in star	I
ASTORIA (Fulton Co.)							
J. M. Todd	Bottle(71)	HB	Aqua				
J. M. Todd & Bro.	Bottle	HB	Aqua		C		
(Reference to J.M.Todd & Bro.)	NBG,9/1886						
J. M. Todd	Bottle(78)	ABM	Aqua	$7\frac{1}{2}$oz.			
AUBURN (Sangamon Co.)							
A.B.W. (Auburn Bottling Works)	Bottle(71)	HB	Aqua		W		
A.B.W. (Auburn Bottling Works)	Bottle(78)	AC	Clear		W		
AURORA (Kane Co.)							
Aurora Bottling Co.	Bottle(78)	AC	Aqua	8 oz.	A		
C & W Bottling Co.	Bottle	ABM-P	Clear	7 oz.	G		G
(Reference to G. W. Green, bottler)	1880 MC						
(J.W. & G.W.Green, bottlers)	1874-75 CD						

Embossed Company &/or Bottler Name. Comments: (Owner's name, date(s) of operation, ownership changes, etc.)	Source of Information (Plate Illus.)	Bottle Style	Bottle Color	Vol.	Major Bottom Marks	Trade Mark	Bottle Manuf.
AURORA (Kane Co.) Cont.							
(John Hall bought bottling business from Anderson and Eckstrom)	NBG, 3/1930						
H. E. Lloyd	Bottle	HB	Clear				
AVISTON (Clinton Co.)							
N. Berns	Bottle(71)	HB, AC	Aqua		NB		
Wm. J. Lager	Bottle(71)	HB	Aqua				
AVON (Fulton Co.)							
Red Avon Spring	Fitch 1927						
BATAVIA (Kane Co.)							
(C.A.Baker)	NBG, 11/1884						
BEARDSTOWN (Cass Co.)							
Wm. Benson & Son	Bottle	ABM	Aqua		B		
Chero-Cola (bottled by Meyers)	Bottle	ABM-P	Clear	6 oz.	Patent pending		
F. W. Ehrhardt	Bottle(34)	TB	Aqua				I G Co.
F. W. Ehrhardt/Superior Belfast Ginger Ale	Bottle(34) (import)	TB	Aqua				
Belfast & Dublin Aromatic Ginger Ale (apparently Ehrhardt)	Bottle(34) (import)	TB	Aqua		E in star		
F. W. Ehrhardt	Bottle(34)	HB	Aqua				I G Co.
W. J. Hagener Bottling Works (some bottles without initials)	Bottle(78)	AC	Aqua				
Meyers	Bottle(30)	ABM-P	Aqua	6 oz.	Star		
Twin-Sodas (Chero-Cola Bottling Co.: bottled by Meyers)	Bottle	ABM-P	Clear	6½oz.	C		
E. E. Rink/Polar Distilled Waters	Bottle(25, 78)	ABM-P	Aqua	7½oz.	R	Polar Bear	
Schneider & Krohe	Bottle	TB	Aqua				
BELLEVILLE (St. Clair Co.)							
L. Abegg	Bottle	TB	Aqua				
(Reference to Louis Abegg)	1860 MC						
Artesian Bottling Works	Bottle	ABM	Aqua	7 oz.			
(August Coob)	NBG, 2/1885						
Culli Bottling Co.	Bottle	ABM-P	Clear	7 oz.	I		I
J. Fisher	Bottle(71)	TB	Aqua		F		A & DHC
(Reference to Joseph Fisher, mineral water bottler)	1880 MC						
G. W. Frees	Bottle	AC	Aqua				
Fred Winkler	Bottle	HB	Aqua				
John Winkler	Bottle	ABM	Aqua	7 oz.			
BELVIDERE (Boone Co.)							
Dooley & Mullen	Bottle	HB	Aqua				
Dooley & Mullen	Bottle	AC	Aqua				SAB Co.
James Walsh	Bottle(71)	HB	Aqua				I G Co.
(James Walsh, ginger ale & soda water bottler)	1894-98 CD						
BENLD (Macoupin Co.)							
Benld Bottling Wks./G.Scopel, Prop.	Bottle(78)	ABM	Clear or Aqua	7 oz.			
Benld Bottling Works	Bottle	ABM-P	Clear	7 oz.	I		I
BENTON (Franklin Co.)							
Benton Ice Cream & Bottling Co.	Bottle	ABM-P	Green	6 oz.	B		Root
BLOOMINGTON (McLean Co.)							
J. B. B.	Bottle	HB	Aqua				I G Co.
Excelsior Bottling Co.	Bottle(71)	HB	Aqua		E.B.Co.		I G Co.
(Some H.blobs with E.B.Co. on bottom only)							
Excelsior Bottling Works/W.F.Schuck	Bottle	AC	Aqua	8 oz.	E.B.W. 8 oz.		
(Reference to Excelsior Bottling Works, W.F.Schuck)	NBG, 12/1901						

Embossed Company &/or Bottler Name. Comments: (Owner's name, date(s) of operation, ownership changes, etc.)	Source of Information (Plate Illus.)	Bottle Style	Bottle Color	Vol.	Major Bottom Marks	Trade Mark	Bottle Manuf.
BLOOMINGTON (McLean Co.) Cont.							
(Excelsior Bottling Works; Edward Holland; Henry & Herman Quosigk /Quosick/ bottlers)	1893, 1902 CD						
E. Holland	Bottle(71)	HB	Aqua				I G Co.
(Edward Holland succeeds Kreis and Bryan)	NBG, 8/1884						
(Edward Holland & William Schuck, bottlers)	NBG, 4/1886						
(E. Holland, soda water bottler)	1885 CD						
J. F. Humphreys & Co./Wishbone Beverages	Bottle	ABM	Aqua	7½oz.		Wishbone	
(Peter Kries, bottler)	1880 MC						
(C.Peterson & Bro., soda & mineral water bottlers)	1873 CD						
H. Quosick	Bottle(25)	HB	Aqua		Anchor	Anchor	Root
H. Quosick	Bottle	ABM	Aqua	7½oz.	Anchor	Anchor	
(Mueller F. Stein, soda water bottler)	1860 MC						
Watkins Fine Beverages	Bottle	ABM	Aqua	7 oz.	W		
Watkins Beverage Co, /Every Swallow Pure	Bottle	ABM-P	Clear or Aqua	7 oz.	W		
BLUE ISLAND (Cook Co.)							
Calumet Bottling Works/Wm.Stein	Bottle	AC	Clear				
G. H. Hausburg	Bottle	HB	Aqua		G.H.H.		A.G.W.
Henry Roll	Bottle	HB	Clear		HR		
Henry Roll	Bottle	AC	Aqua			S.D.	Root
Wm. Roll	Bottle	ABM	Clear	6½oz.			
William Roll	Bottle	ABM	Aqua	7 oz.	W.R.	S.D.	
Roll's Beverages	Bottle	ABM-P	Clear	7½oz.	Blue Island Ill.		G
Wm. Stein	Bottle	HB	Aqua		Star		C & Co Lim.
BRACEVILLE (Grundy Co.)							
Allen & Son	Bottle	ABM	Aqua	7 oz.	75		
Braceville Bottling Works	Bottle	AC	Aqua		75		
BRAIDWOOD (Grundy Co.)							
Blood's Bottling House	Bottle	HB	Aqua				T.W. & Co.
H. W. Blood & Co.	Bottle	HB	Aqua				
(Business started in Wilmington.H.W. Blood & Son successors to H.W.Blood)	NBG, 6/1891						
Jas. Hynd(or) James Hynd	Bottle(78)	AC	Aqua		H	JH	C & Co.; C & Co., Lim.
Hynd	Bottle	ABM	Clear & Aqua		H		
Hynd	Bottle	ABM-P	Clear	7 oz.	H		
Hynds Beverages	Bottle	ABM-P	Aqua	7 oz.	I 7 fl.oz.		I
(Hynd business started ca.1888: letter from Bernice Hynd 10/22/71)							
BREEZE (Clinton Co.)							
C. Dorries & Brother	Bottle	TB	Aqua				Wm.McC & Co.
C. Dorries	Bottle	ABM	Clear				
Excel Bottling Co.	Bottle(29)	ABM-P	Clear	7 oz.	XL		N
Prairie View Dairy/D-Lish-O	Bottle	ABM-P	Clear	7½oz.	Cap.7½ FLU.OZS.		
BRIDGEPORT(Lawrence Co.)							
The Crystal Ice & Supply Co.	Bottle	ABM	Aqua				
BRISTOL (Kendall Co.)							
(G.G.Hunt erected new bottling plant)	NBG, 1/1896						
Min-Ni-Yan Springs	Fitch 1927						
Mini-Niyan Spring Water Co.	Bottle	AC	Aqua	1pt.8oz.	AB Co.		AB Co.

Embossed Company &/or Bottler Name. Comments: (Owner's name, date(s) of operation, ownership changes, etc.)	Source of Information (Plate Illus.)	Bottle Style	Bottle Color	Vol.	Major Bottom Marks	Trade Mark	Bottle Manuf.
BROOKFIELD (Cook Co.)							
Brookfield Bottling Co.	Bottle	ABM	Aqua	6½oz.	BBC		
BUSHNELL (McDonough Co.)							
Bushnell Bottling Works	Bottle(78)	ABM	Aqua	8 oz.	M		
Bushnell Bottling Works/S.O., Prop.	Bottle(71)	HB	Aqua				I G Co.
New England Bottling Works/S.O., Prop.	Bottle	AC	Clear		M		
CAIRO (Alexander Co.)							
H. Hasenjaeger	Bottle(71)	HB	Aqua		15		
(Reference to Henry Hasenjaeger)	NBG, 8/1891						
Andrew Lohr Bottling Co.	Bottle	HB	Aqua				C. & CO. Lim.
Andrew Lohr Bottling Co.	Bottle(78)	AC	Aqua			Bottle/L in wreath	C. & CO. Lim.
Andrew Lohr Bottling Co.	Bottle(78)	ABM	Clear	6½oz.	Bottle/L in wreath	Bottle/L in wreath	
(Reference to the Lohr Company)	NBG, 5/1891 2/1892						
CALUMET CITY (Cook Co.)							
(Northern Beverage Co.)	NBG, 7/1925						
CANTON (Fulton Co.)							
(Reference to George Bryan)	1880 MC						
(G. A. Bryan)	NBG, 8/1884						
Rene Butel	Bottle	AC	Aqua		Star		I G Co.
Canton Bottling Works	Bottle	HB	Aqua		16		
Canton Bottling Works	Bottle	AC	Clear				I G Co.
Canton Bottling Works	Bottle	ABM	Aqua	6½oz.			
(C.L.Harvey sold bottling business to C.E.Clemens of Galesburg)	NBG, 11/1899						
Miller Bottling Works	Bottle(29, 78)	ABM-P	Clear	6½oz.			I
Miller Beverage Co.	Bottle(30)	ABM-P	Green	6 oz.	Canton, Ill.		I
(J.L.Westervelt succeeded the firm of George Bryan & Son)	NBG, 1/1892						
CARBONDALE (Jackson Co.)							
(Carbondale Bottling & Ice Cream Co.: Terpinitz & Co.)	NBG, 5/1910						
CARLINVILLE (Macoupin Co.)							
(Chas. Labbe & Son, bottlers)	1880 MC						
Sanitary Bottling Works	Bottle	ABM	Clear	6½oz.	GOOCH		
H. Segelke & Co.	Bottle(71)	HB	Aqua				
John Weber	Bottle(71)	TB	Aqua				A & DHC
Weber & Mueller	Bottle	TB, HB	Aqua				A & DHC
Aug. Zaepffel	Bottle(72)	HB	Aqua				I G Co.
CARLYLE (Clinton Co.)							
Carlyle Bottling Works	Bottle	ABM	Clear	7 oz.			
Carlyle Bott. Wks./Supreme Beverages	Bottle	ABM-P	Clear	7 oz.	Conts.7 FLU.OZS.		
H. Hess & Co.	Bottle	TB	Aqua				
J. H. Menkhaus	Bottle	HB	Aqua				
CARMI (White Co.)							
(Louis Fauss, mineral water bottler)	1880 MC						
CARROLLTON (Greene Co.)							
H. Schafer, Prop.	Bottle	ABM	Clear	6½oz.			
(Carrollton Bottling Works, H.Schafer, Prop., sold to Hartman & Berlin)	NBG, 8/1925						
CARY (McHenry Co.)							
Minerva Mineral Spring	Fitch 1927						
CAVE IN ROCK (Hardin Co.)							
Cave-in-Rock Bottling Works	Bottle	ABM	Aqua	6½oz.			

Embossed Company &/or Bottler Name. Comments: (Owner's name, date(s) of operation, ownership changes, etc.)	Source of Information (Plate Illus.)	Bottle Style	Bottle Color	Vol.	Major Bottom Marks	Trade Mark	Bottle Manuf.
CENTRALIA (Marion Co.)							
Centralia Bottling Works	Bottle(78)	AC, ABM	Aqua	8 oz.			
Centralia Bottling Works/Pure Food Soda Water	Bottle	ABM	Aqua	6½oz.	CWB		
(Centralia Cider Mills)	NBG, 5/1891						
Mathis & Davidson	Bottle	HB, AC	Aqua				
Robert Rohl	Bottle	HB	Aqua				NBBG Co.
(Reference to Robert Rohl)	NBG, 2/1896						
(Robert Rohl bought out Ed Hayes)	NBG, 7/1886						
A. J. Sligar Bottling Works	Bottle	AC	Aqua				
The Union Bottling Works	Bottle(78)	ABM	Aqua		U		
(Reference to the Union Bottling Works)	NBG, 3/1915						
Woolner & Co.	Bottle(72)	HB	Aqua				NBBG Co.
CHAMPAIGN (Champaign Co.)							
(Champaign Bottling Works)	NBG, 4/1886						
Champaign Bottling Works	Bottle(78)	Semi-ABM	Green		M.M.		Root
(Reference to N. Miller, mineral water bottler)	1880 MC						
(Schumacher & Shore)	NBG, 12/1884						
CHARLESTON (Coles Co.)							
Charleston Bottling Works	Bottle(72)	HB	Aqua		33		NBBG Co.
Galbreath & McKinney	Bottle	ABM	Aqua	6½oz.			
Walton & Schriner Bottling Works	Bottle	ABM	Aqua				
CHESTER (Randolph Co.)							
Tivoli Spring	Fitch 1927						
C. Welge & Son	NBG, 3/1891						
John Wenda	Bottle(72)	TB	Aqua				A & DHC
CHICAGO (Cook Co.)							
(Aerox Distilled Water Co. beginning operations)	NBG, 2/1926						
(Allouez Mineral Spring Co.)	1902 CD						
(Almananis Co.)	NBG, 8/1891						
American Home Brand/Pure Beverages (Geo. Rasmussen Co.)	Bottle	ABM	Clear	1 pt., 8 oz.			
(American Spring Water Supply Co.)	1902 CD						
(Andrews Magnetic Mineral Spring Water)	1902 CD						
(P. I. Athenson, mineral water bottler)	1870 MC						
(Barothy & Gilman, bottlers)	Pers. comm. Chas. Lorenz, Chicago, 9/29/70						
(Charles Benzing, mineral water bottler)	1902 CD						
(M. Bernstein & Co., bottler)	1902 CD						
(Bethesda Mineral Spring Co.)	1902 CD						
Big Ben Beverage Corp.	Bottle	ABM-P	Clear	8 oz.	Big Ben Beverage Corp. Chicago, Ill.		
(George Bisbro & Co.)	NBG, 2/1885						
G.A. Bode/Bottlers Extracts	Bottle	HB	Aqua				A.G.W.
Bode/No.5	Bottle	HB	Clear				
Bon-Ton	Bottle	ABM-P	Aqua	7 oz.	Mulvihill Good Wine Company		
D. W. Bostelmann	Bottle	HB	Aqua		Anchor	Anchor	
D. W. Bostelmann	Bottle	HB	Aqua	quart	C. & CO. LIM.	Anchor	C & CO., Lim.
Bostelmanns	Bottle	ABM	Aqua	1 pt., 8 oz.	D. W. Bostelmann Co., Chicago, Ill. Anchor	Anchor	

Embossed Company &/or Bottler Name. Comments: (Owner's name, date(s) of operation, ownership changes, etc.)	Source of Information (Plate Illus.)	Bottle Style	Bottle Color	Vol.	Major Bottom Marks	Trade Mark	Bottle Manuf.
CHICAGO (Cook Co.) Cont.							
(James A. Bowman, mineral water bottler)	1902 CD						
(C. M. Boyce)	NBG, 1/1892						
Bristol Mineral Spring Co.	Bottle(72)	HB	Aqua			BB	W.F.&S. MIL.
Henry Burkhardt	Bottle	HB	Aqua		HB		
Henry Burkhardt	Bottle	AC	Aqua				NBBG Co.
Henry Burkhardt Beverages/since 1886	Bottle	ABM-P	Clear	8 oz.& 10 oz.	G		G
Henry Burkhardt's Beverages	Bottle	ABM	Aqua				
C.M.W.Co.	Bottle(72)	HB	Aqua		WIS.GLASS CO., MILK.		Wis.Glass Co.
California Beverage Co.	Bottle	ABM-P	Aqua	7 oz.			C.B.C.
Canfield's	Bottle	ABM-P	Aqua	8 oz.			A.J.Canfield Co. Chicago, Ill.
Carbonated Beverage Works/ Columbia Brand	Bottle	ABM	Aqua	1 pt., 8 oz.	CBW	CBW	
A. Carpenter	Bottle(72)	HB	Aqua			AC	I G Co.
20th Century Bottling Works	Bottle	ABM	Clear	1 pt., 8 oz.	I.L.	Lion	
Chicago Beverage Co./De Luxe	Bottle	ABM	Aqua		Purity, Virtue, Merit (in a Y)		
Chicago Consolidated Bottling Co.	Bottle	HB	Aqua			CCB Co.	E.H.E.CO., N.B.B.G.Co. or A.G.W.L.
Chicago Consolidated Bottling Co. (George Lomax, Pres.)	Bottle	Semi-ABM	Aqua		ROOT	CCB Co.	Root
Chicago Consolidated Bottling Co./ Splits/George Lomax, Pres.	Bottle	AC	Aqua			CCB Co.	A B Co., Root NBBG Co.
Chicago Extract Mfg. Co.	Bottle	HB	Aqua				Root
(Chicago Fountain Soda Water Co.)	NBG, 6/1896						
(Chicago Pure Beverage Co.)	NBG, 11/1891						
Chicago White Rose Bottling Works	Bottle	AC	Aqua			Rose	A B Co.
(Chippewa Spring Water Co.)	NBG, 11/1891						
Arthur Christin	Bottle	HB	Aqua		A.C.		D.O.C.
(Arthur Christin, ginger ale bottler)	1880 MC						
Arthur C∠h7ristin	Bottle	TB	Cobalt				A & DHC
Ciovino and Arata	Bottle	HB	Aqua				
(Colorado Lithia Water Co.: agent for Pluto Water)	1902 CD						
(Columbia Bottling Co.)	NBG, 11/1891						
(Columbia Mineral Water Co.)	NBG, 7/1891						
The Consumers Co. (Hydrox)	Bottle(24)	TB	Aqua	½ gal.	Hydrox	Hydrox	A B Co.
(Edward Cronk, Sarsaparilla bottler)	1880 MC						
(Crystal Water Co.)	1902 CD						
D & H/Ginger Ale	Bottle	HB	Aqua		D & H		
(Dallemand & Co.)	NBG, 5/1891						
(Deer Lick Mineral Spring Co.)	1902 CD						
(Delavan Springs Co.)	NBG, 8/1891						
(C. Del Genio & Company)	NBG, 6/1896						
D. Dempsey (South Chicago)	Bottle	HB	Aqua				M.G.CO.
J. Dinets Superior Soda Water	Schmeiser (1970)						
(George Dishro & Co.)	NBG, 8/1886						
V. Donat	Bottle	HB	Aqua & Clear		D		
(Donat & Kara successors to V. Donat)	NBG, 8/1891						
Donat & Motis	Bottle	HB	Aqua		D		NBBG Co.

Embossed Company &/or Bottler Name. Comments: (Owner's name, date(s) of operation, ownership changes, etc.)	Source of Information (Plate Illus.)	Bottle Style	Bottle Color	Vol.	Major Bottom Marks	Trade Mark	Bottle Manuf.
CHICAGO (Cook Co.) Cont.							
Eastern Bottling Works	Bottle	HB	Aqua		JS	Steam-ship	
G. Eberlein	Bottle	TB	Green				
(Albert Ellithorpe, mineral water bottler)	1902 CD						
Extract Mfg. Co.	Bottle	HB	Aqua				Root
Failing & Wagner Bottling Works	Bottle	HB	Aqua		F & W		
(Failing & Wagner, soda bottlers)	1902 CD						
(Fairbanks Beverage Co., just organ-ized)	NBG, 5/1925						
Federal Beverage Co./High Life	Bottle	ABM	Aqua	7 oz.			
Feldman Bottling Works	Bottle	HB	Aqua			PF	
Five-O Dist. Co.	Bottle(29)	ABM-P	Clear	6½oz.	5 in circle	Five-O	
(Anthony Fredricks, bottler)	1902 CD						
(Frisch & Co., mineral waters)	1902 CD						
(Ernst Funk)	NBG, 9/1886						
(Hunt Galpaugh, soda water bottler)	1870 MC						
(Henry Garben, mineral water bottler)	1902 CD						
Garden City Bottling Works	Bottle(24)	ABM	Aqua	8 oz & 1 pt., 9 oz.		RS	
Shubal Geen, pop beer bottler)	1880 MC						
(Otto K. Gernsmeyer, soda & mineral water bottler)	1870 MC						
(Gettysburg Mineral Springs Co.)	1902 CD						
Globe Beverage Co.	Bottle	ABM	Aqua	7 oz.	Globe	Globe	
(Golden Health Co., just organized)	NBG, 5/1925						
Goodwine Bottling Co.	Bottle(24)	ABM	Aqua	1 pt., 8 oz.	G	Rooster	
(Thomas Gregg, mineral water bottler)	1902 CD						
Gregg's Wyaconda Diuretic Mineral Water	Bottle	TB	Aqua	½gal.?			
The Haas Co.	Bottle	HB	Aqua		THC	THC	
(Wm. Hafner, soda water manuf.)	1870 MC						
Hafner & Will/Mfrs. Soda Water Machinery	Bottle	HB	Aqua		D.O.C.		D.O.C.
(Harrison & Bostelman, new bottlers)	NBG, 5/1896						
Hass Bros.	Bottle	HB	Aqua		H		
Chas. G. Hausburg	Bottle(24)	ABM	Aqua	1 pt., 8 oz.	CGH	CGH	
Wm. A. Hausburg	Bottle	HB	Aqua				
Hayes Bros.	Bottle	AC	Aqua				C & Co., A B Co.
Hayes Bros.	Bottle	ABM	Aqua	1 pt., 8 oz.	ROOT	Horseshoe	Root
Hayes Bros. (Established 1871)	Bottle(25)	HB	Aqua			Horseshoe	A.G.W.
(Reference to Hayes Bros.)	1880 MC						
Hayes Bottling Co.	Bottle	HB	Aqua		7	F	
Hays Bros.	Bottle	TB	Aqua				A & DHC
(Henk Mineral Spring Water Co.)	1902 CD						
Thomas Hennessy	Bottle	HB	Aqua		TH		I G Co.
(Reference to Dempsey & Hennessy, bottlers)	1880 MC						
John Hessler Bottling Works	Bottle(72)	HB	Aqua		JH	Star	
Frank Higgins	Bottle	HB	Aqua		FH		
(Frank J. Higgins, bottler)	1902 CD						
(O. W. Hinckley, mineral waters)	1902 CD						
Hinckley & Schmidt	Bottle(69)	ABM	Aqua	1 gal.	I		I
Jos. Hladovec	Bottle	HB	Aqua				
(Honneberg & Sampson)	NBG, 1/1892						

Embossed Company &/or Bottler Name. Comments: (Owner's name, date(s) of operation, ownership changes, etc.)	Source of Information (Plate Illus.)	Bottle Style	Bottle Color	Vol.	Major Bottom Marks	Trade Mark	Bottle Manuf.
CHICAGO (Cook Co.) Cont.							
R. L. Hunt	Bottle	TB	Aqua				FA & Co.
(Reference to W.H.Hutchinson)	1860-1880 MC						
W. H. H. (Hutchinson)	Bottle	HB	Cobalt				
W. H. Hutchinson & Son	Bottle	AC	Aqua	quart?			
Hydrox	Bottle	ABM-P	Aqua	6½oz.			I
Ideal Beverage Co.	Bottle	ABM	Aqua	1 pt., 8 oz.			
Independent Beverage Co.	Bottle	ABM-P	Aqua	7 oz.	Contents 7 Fl.oz.		I
(Illinois Perfect Condensation Co.)	1902 CD						
(Indellino Mineral Water Co.)	NBG, 8/1891						
(Ishams California Waters of Life)	1902 CD						
Kantor Bros. & Co.	Bottle	HB	Aqua		KB	Steer head	SB & G Co.
(Mayer Kanter, mineral water bottler)	1902 CD						
M. Keeley	Schmeiser 1970						
(M. Keeley, soda and ale bottler)	1860-1870 MC						
King & Joyce	Bottle	HB	Aqua		K & J		Root
Charles Kluetsch	Bottle	HB	Aqua		CK TRADE MARK	CK	C & Co. Lim.
Kosciuszko Bottling Works	Bottle	HB	Aqua				
Ladroc Beverages (Jewell Water Co.)	Bottle	ABM-P	Green	7½oz.	I		I
Lakeside Bottling Co.	Bottle	TB	Amber	1 pt., 8 oz.			
Lamb Beverage Co.	Bottle	AMB-P	Aqua	7½fl.oz.	LAMB	Lamb	
(Lang Bros., mineral water bottlers)	1870 MC						
F. C. Lang	Bottle	HB	Aqua		PUTNAM 6		D.O.C.
F. C. Lang Bottling Co.	Bottle	HB	Aqua		FCL		C & Co., Lim.
(Reference to F.C.Lang, bottler)	1880 MC						
(John Lasser, mineral water bottler)	1902 CD						
(Harry E. Lazarus, bottler)	1902 CD						
M. Levin & Son	Bottle	HB	Aqua		M.L.	Elephant	
I. Levinson	Bottle(25, 72)	HB	Aqua		I.L.	Reclining lion	C & Co., Lim. I G Co.
(Harry Levitan, bottler)	1902 CD						
(Frederick Lewis, mineral water bottler)	1902 CD						
Liberty Bottling Co.	Bottle	ABM	Clear	7 oz.	Statue of Liberty	Liberty	
Liberty Bottling Co.	Bottle	ABM	Aqua	24 oz.	Liberty	Statue of Liberty	
Liberty Bottling Co. (Blue Rock Beverages)	Bottle(24)	ABM	Aqua	1 pt., 8 oz.	Statue of Liberty	Statue of Liberty	
Lincoln Bott. Co.	Bottle	ABM-P	Clear	7½oz.	LB CO		
(Lincoln Mineral Water Co.)	1902 CD						
G. Lomax (ca. 1850-59)	Bottle	TB	Green & Cobalt				
J. A. Lomax	Bottle	HB	Cobalt		J.L.	JAL	
J. A. Lomax/Ginger Ale	Bottle	HB	Aqua				A & DHC
(Reference to John A. Lomax)	1880 MC						
Lomax's Beverages	Bottle	AC	Aqua		ROOT		Root
Lomax's Chicago Beverages	Bottle	ABM	Aqua	24 oz.			

Embossed Company &/or Bottler Name. Comments: (Owner's name, date(s) of operation, ownership changes, etc.)	Source of Information (Plate Illus.)	Bottle Style	Bottle Color	Vol.	Major Bottom Marks	Trade Mark	Bottle Manuf.
CHICAGO (Cook Co.) Cont.							
The Lomax Co.	Bottle	ABM	Aqua	6½oz.			
Lorraine Beverages	Bottle	ABM-P	Clear	8 oz.	Lorraine Bev. Co., Chicago, Ill.		G
Manhattan Beverages	Bottle(24)	ABM	Aqua	1 pt. 9 oz.	Horse head	Horse head	
Manhattan Bottling Co.	Bottle(25)	HB	Aqua		Trademark (horse head) Registered	Horse head	G & Co. Lim.
Manhattan Bottling Co.	Bottle	AC	Aqua			Horse head	I G Co.
Manhattan Bottling Co.	Bottle	ABM	Aqua	7 oz.	Horse head	Horse head	
(Frank Matejka, bottler)	1902 CD						
Matejka & Gruener	Bottle	HB	Aqua		Star	Star	EHE Co.
(Mathews Soda Water Co.)	NBG, 11/1891						
(Metropole Bottling Co., new firm)	NBG, 6/1896						
Mette Bros.	Bottle	HB	Aqua				
(A. Mette & Bro.)	NBG, 6/1884						
A. Mette & Bro.	Bottle	HB	Aqua		M. Bros.		McC
(Reference to A. Mette & Bros.)	1880 MC						
(Floyd Miller, mineral water bottler)	1902 CD						
Miller & Kluetch	Bottle	HB	Aqua		M & K		C& Co. Lim.
Modern Beverage Co.	Bottle	ABM	Aqua				
Modern Perfection Bottling Co., Inc.	Bottle	ABM-P	Aqua	7 oz.			
Monarch Beverage Co.	Bottle	ABM	Aqua	7 oz.			
Morand Bros.	Bottle	TB	Aqua		A.G.W.L.		A.G.W.L.
P. O. Morand Bros.	Bottle	AC	Amber				
(Abraham Mottlowitz, bottler)	1902 CD						
Mulvihill - Bell Bot. Wks.	Bottle	ABM	Aqua	1 pt., 8 oz.	M.B.B.Co. in a bell	Bell	Root
J. N. Mulvihill	Bottle	HB	Aqua		M		
(Mrs. Mary Mulvihill, bottler)	1902 CD						
Wm. Mulvihill	Bottle	AC	Aqua				R G CO
(William Mulvihill, bottler)	1902 CD						
Mulvihill's Beverages	Bottle	ABM-P	Clear	7 oz.	MC	MC	I
Jos. Najemnik	Bottle(25)	HB	Aqua		JN	Lion	N.B.B.G.Co.
Najemnik & Vana	Bottle	HB	Aqua		N & V	Lion	S.B.& G Co.
Oakley Bottling Works	Bottle	ABM	Aqua	7 oz.	OBW		
(Olympian Springs Water Co.)	NBG, 8/1891						
(Otsego Mineral Water Co.)	NBG, 11/1891						
(F. H. Peabody & Co., bottlers)	Pers. Comm. Chas. Lorenz, Chicago. 9/29/70						
(People's Distilled Water Co. incorporated)	NBG, 12/1901						
Perfection Bottling Works	Bottle	HB	Aqua		P.B.W.		A B Co
J. B. Peto & Co, /Triangle Beverages	Bottle	ABM	Clear				
Polonia Bottling Works	Bottle (79)	ABM	Aqua	7 oz.		Man on horseback	I
Pride Bottling Co.	Bottle	HB	Aqua		P.B.	Elk	
(Reference to Pride Bottling Co.)	NBG, 10/1910						
M. Puchowich Co.	Bottle(25)	HB	Aqua		M.P.	Bird	
(A. D. Puffer & Sons)	NBG, 4/1891						
(Purex Mineral Spring Co.)	1902 CD						
(Rascusin and Michelson, new firm)	NBG, 11/1905						
Geo. Rasmussen Co.	Bottle	ABM	Aqua				Root
(Ratto Soda Water Co.)	NBG, 11/1891						
(Paul Rauze, bottler)	Pers. Comm. Chas. Lorenz, Chicago, 9/29/70						
(Reid, Murdock & Fischer)	NBG, 7/1886						
H. Rocatz	Bottle	HB	Aqua		28	HR	

Embossed Company &/or Bottler Name. Comments: (Owner's name, date(s) of operation, ownership changes, etc.)	Source of Information (Plate Illus.)	Bottle Style	Bottle Color	Vol.	Major Bottom Marks	Trade Mark	Bottle Manuf.
CHICAGO (Cook Co.) Cont.							
L. Rodemeyer & Co./Premium Mineral Water	Bottle	TB	Green				
Rusch Bottling Co.	Bottle	ABM-P	Aqua	½ pint	Full ½ pint		
Ryan Bottling Works	Bottle(72)	HB	Aqua		R.B.W.	R.B.W.	
Ryan Bottling Works	Bottle	ABM	Aqua		R.B.W.	R.B.W.	
Ryans	Bottle	ABM	Aqua	7 oz.	RYANS	RYANS	
(Sangcura Sprudel Wells Co.)	1902 CD						
L. H. Sass	Bottle	HB	Aqua		O		C.&Co. Lim.
Sass & Pomy	Bottle	HB	Aqua		O		
(Seligman Bros., mineral water bottlers)	1902 CD						
(Shure Bros, soda water bottler)	1870 MC						
Singer & Rosenbaum	Bottle	HB	Aqua			Deer ?	
South Chicago Bottling Works	Bottle	HB	Aqua		Star		NBBG Co.
South Chicago Bottling Works	Bottle(24)	ABM	Aqua	7 oz. & 24 oz.	LW	LW	
M. B. Spatz F. S. Co.	Bottle	HB	Aqua				
Spring Beverage Co.	Bottle	ABM-P	Aqua	7 oz.		Tulip & Leaves	
Spring Beverages	Bottle	ABM	Clear	8 oz.	Spring Bev. Co.Chicago		
Star Bottling Works	Bottle	AC	Clear		Star	Star	
James Stenson	Bottle	TB	Aqua				A & DHC
Jas. Stenson	Bottle	HB	Aqua		D.O.C.		D.O.C.
(Reference to James Stenson)	1880 MC						
Sterling Beverage Co.	Bottle	ABM-P	Clear	8 oz.			
Success	Bottle	ABM-P	Clear		Contents 8 Fl. oz.Chicago, Ill.		I
Success Bottling Works	Bottle	AC	Clear				
Success Bottling Works	Bottle(25)	HB	Clear			Pig	
Success Bottling Works	Bottle	ABM	Aqua	7 oz.		Pig	
Success Bottling Works	Bottle	ABM	Aqua	7 oz.	Horse		
Sunbeam Beverage Co.	Bottle(24)	ABM	Aqua	1 pt., 8 oz.	S. & B.	Sunburst	
Supreme Beverage Co.	Bottle	ABM	Aqua	7 oz.			
Supreme Beverage Co./A Supreme Product	Bottle	ABM	Aqua	7 oz.	Supreme		
Swidler & Bernstein (Sunbeam, and Swidler & Bernstein bottles bear the same address: 1527-1529 Flournoy St.)	Bottle(25,72)	HB	Aqua	8 oz.	S. & B.	Bear	
Sylvan Dell Sulpho-Magnesian Spring	Fitch 1927						
(Herman Tessendorff, bottler)	1902 CD						
(Trevor Spring Water Co.)	1902 CD						
Uncle Sam Beverage Co.	Bottle	ABM-P	Aqua	7 oz.	U.S.	U.S.	
Union Protective Bottling Works	Bottle	HB	Aqua				I.G.Co.
United Carbonating Co.	Bottle	ABM	Aqua	7 oz.			
Alois Vana	Bottle	HB	Clear		A.V.	Lion	
Alois Vana (Previously Najemnik & Vana: 82-84 Fisk St.)	Bottle	ABM	Aqua	6½ & 7 oz.	A.V.	Lion	
Vana Inc.	Bottle	ABM-P	Clear	8 oz.	G		G
(Vance Bottling Co.)	NBG, 8/1891						
(Paul Vandenberg, bottler)	1902 CD						
Victoria Bottling Works/S.Stein & Son	Bottle	HB	Aqua		S & S		
Henry Weaber	Bottle	HB	Aqua		W & D		
(Henry Weber)	NBG, 2/1892						

Embossed Company &/or Bottler Name. Comments: (Owner's name, date(s) of operation, ownership changes, etc.)	Source of Information (Plate Illus.)	Bottle Style	Bottle Color	Vol.	Major Bottom Marks	Trade Mark	Bottle Manuf.
CHICAGO (Cook Co.) Cont.							
(Weaber and Darley, New Business)	NBG, 4/1884						
Weaber & Darley	Bottle	HB	Aqua		WD	WD	
A. Wember	Bottle(72)	HB	Aqua				NBBG Co.
(John Wendt, bottler)	1902 CD						
Chas. Westerholm & Co.	Bottle(72)	HB	Aqua		C W & Co	C W Co.	NBBG Co.
Chas. Westerholm & Co.	Bottle	AC	Aqua			C W Co.	Root
Western Bottling Works	Bottle	HB	Aqua		Trade Mark (clasped hands) Registered	clasped hands	C & Co. Lim.
West Side Bottling Works	Bottle	HB	Aqua		Trade Mark SM. Reg.		A B Co
White Eagle Bottling Co. (Bibo)	Bottle	ABM	Aqua	24 oz.	P & R	BIBO	
(White Rock Mineral Spring Co.)	1902 CD						
Wolf Wirchowski	Bottle(25)	HB	Aqua			Wolf	
(G. Wurater, bottler)	1880 MC						
(G. Wurater)	NBG, 8/1886						
Wurster & Co.	Bottle	TB	Aqua		W. & Co.		A & DHC
Yankee Bottling Works	Bottle	ABM-P	Aqua	7 oz.	Y	Man beating a drum	
Zajicek Bros.	Bottle	HB	Aqua		29	Rabbit	
(Zajicek Bros., bottlers)	1902 CD						
CHILLICOTHE (Peoria Co.)							
Crutchfield Brothers	Bottle	ABM-P	Green & Clear	6½oz.	C	Indian head	Root
Jno. R. Gilfillan	Bottle	ABM	Aqua	7 oz.	C		
(John Gilfillian purchased soda pop factory from Tony Anderson)	NBG, 10/1909						
Tony Anderson	Bottle	HB	Aqua				
CICERO (Cook Co.)							
Cicero Bottling Works	Bottle	ABM-P	Clear	7½oz.	Cicero		I
Cicero Bottling Works	Bottle	ABM	Aqua	1 pt., 8 oz.	Cicero		
Hawthorn Beverage Works	Bottle	ABM-P	Aqua	7 oz.	G		G
Hawthorn Beverage Works/We Cure Your Thirst	Bottle	ABM-P	Clear	7 oz.	Cicero Ill.		I
CLAYTON (Adams Co.)							
Clayton Bottling Co.	Bottle	ABM	Aqua		C		
Clayton Bottling Co.	Bottle(79)	ABM	Clear	7 oz.			
CLINTON (DeWitt Co.)							
Clinton Bottling Works	Bottle	ABM	Aqua		C		
COLLINSVILLE (Madison Co.)							
H. Baierlein	Bottle	TB	Aqua				
B. Bischoff	Bottle	HB	Aqua		BB		AGWL
Bischoff Bottling Works	Bottle(72)	HB	Aqua				
COLUMBIA (Monroe Co.)							
John Gundlach	Bottle	TB	Aqua		J.G.		L G Co.
Edward Gundlach	Bottle	AC	Aqua				
CREAL SPRINGS (Williamson Co.)							
Creal Springs Bottling Co.	Bottle	ABM-P	Clear	6½oz.	Creal Springs, Ill.		
Creal Springs	Fitch 1927						
DALLAS CITY (Hancock Co.)							
Crystal Springs Water Company	Bottle	AC	Aqua		SC		
DANVILLE (Vermilion Co.)							
Hudson Brown	Bottle	HB	Aqua				
H. F. Depke Bottler	Bottle(24)	HB	Aqua	quart			E.H.E.
H. F. Depke Bottler	Bottle	AC	Aqua		H.F.D.		
Dudenhofer Bottling Co.	Bottle	AC	Aqua		S in circle		
(C.J. Leverenz to open new bottling works)	NBG, 11/1899						

Embossed Company &/or Bottler Name. Comments: (Owner's name, date(s) of operation, ownership changes, etc.)	Source of Information (Plate Illus.)	Bottle Style	Bottle Color	Vol.	Major Bottom Marks	Trade Mark	Bottle Manuf.
DANVILLE (Vermilion Co.) Cont.							
Leverenz & Brown Bottling Co.	Bottle	HB	Aqua		L.&B.B.Co.	LBB Co	
Peter Perl	Bottle	HB	Aqua				E.H.E.Co.
Reget & Britt Bottling Co.	Bottle	HB	Aqua		R & B		
Reget & Britt Bottling Co.	Bottle(79)	AC	Clear & Aqua		R. & B.		Root
Chas. Shean	Bottle	HB	Aqua				
(Charles Shean succeeds Peter Perl)	NBG, 2/1896						
DARMSTADT (St. Clair Co.)							
Gross & Fleischman	Bottle(72)	TB	Aqua				A & DHC
DECATUR (Macon Co.)							
Gus. Ahrens	Bottle(79)	AC	Aqua		G.A.		Root
(Gus Ahrens prop. of Kuney Bottling Works)	NBG, 1/1896						
(Gus Ahrens sold to Pabst Brew.Co., Milw., but continued bottling under same name)	NBG, 9/1896						
Carburetted Springs	Fitch 1927						
Damrow Bros.	Bottle	HB	Aqua		DB		
(Reference to installation of equipment)	NBG, 4/1896						
Decatur Bottling Works	Bottle(72)	HB	Aqua		D B Co in diamond		C.G.Co.
Decatur Bottling Works	Bottle(24)	AC	Aqua	quart	D B Co in diamond		EHE CO
Decatur Bottling Works	Bottle(79)	AC	Aqua		D B Co in diamond		
C. P. Friis/Grape Foam	Bottle(79)	AC	Amber		C.P.FRIIS & CO.DECATUR, ILLS.		NBBG CO
F. Kuny	Bottle(72)	TB	Green				
(Reference to Fred Kuny Bottling Works)	1880 MC						
Union Bottling Works	Bottle(72)	HB	Aqua		KRAFT		A B Co.
DEERFIELD (Lake Co.)							
Deerlick Spring	Fitch 1927						
DeKALB (DeKalb Co.)							
J.W.Corkings sold bottling business to Henry Hoyt)	NBG, 3/1901						
DePUE (Bureau Co.)							
DePue Bottling Works	Bottle	AC	Aqua	7 oz.			
DIVERNON (Sangamon Co.)							
Divernon Bottling Works	Bottle	AC	Clear				
DIXON (Lee Co.)							
Blue Label Bottling Works of D.W.T.Co.	Bottle(79)	ABM	Aqua	7 oz.	D		
John Fellows	Bottle(79)	ABM	Clear & Aqua	7 oz.			
(John Fellows sold bottling works to Gustave L. Goob)	NBG, 4/1910						
(Henry Wahl bottling business sold to Arthur Gumtan)	NBG, 12/1904						
DIXON SPRINGS (Pope Co.)							
Dixon Springs	Fitch 1927						
Na-Mo-Na Spring	Fitch 1927						
Old Ironsides	Fitch 1927						
DUPO (St. Clair Co.)							
L. J. Parrin Bottling Works	Bottle(79)	ABM	Aqua		P		
DuQUOIN (Perry Co.)							
(Bottling business incorporated by Wm.Henke, W.R.Hayes and Geo. A.Sanford)	NBG, 11/1905						
DuQuoin Bottling Co.	Bottle(79)	AC	Aqua				

Embossed Company &/or Bottler Name. Comments: (Owner's name, date(s) of operation, ownership changes, etc.)	Source of Information (Plate Illus.)	Bottle Style	Bottle Color	Vol.	Major Bottom Marks	Trade Mark	Bottle Manuf.
DuQUOIN (Perry Co.) Cont.							
DuQuoin Bottling Co.	Bottle	ABM	Aqua	8 oz.			
DuQuoin Bottling Co.(Midwest Quality Beverages)	Bottle(29)	ABM-P	Aqua	6 oz.	DUQUOIN ILL.		
DuQuoin Bottling Co.(Quality Beverages)	Bottle(79)	ABM	Aqua	6 oz.			
DWIGHT (Livingston Co.)							
W & W Ice Cream & Chero Cola Co.	Bottle	ABM	Aqua	7 oz.	W & W		
EAST DUBUQUE (Jo Daviess Co.)							
(M. Hollenfelz)	NBG, 10/1896						
EAST DUNDEE (Kane Co.)							
(Miller Beverage Co.)	NBG, 10/1925						
EAST ST. LOUIS (St. Clair Co.)							
E. Auer	Bottle	HB	Aqua		A		
(Emanual Auer & William Jo, bottlers)	NBG, 8/1891						
City Bottling Wks.	Bottle(79)	AC	Aqua	7 oz.	C		
Charles Hufschmidt	Bottle(79)	AC	Clear		H		
Kenwood Club Bottling Co.	Bottle	ABM-P	Clear	7 oz.	56		
(Christ. Lutt & Co.)	1880 MC						
Phos-Ferrone Bottling Works	Bottle	AC	Clear		P.F.		
Phos-Ferrone Bottling Works	Bottle	ABM	Aqua		P.F.		
Thos. Schoenbein & Co.	Bottle	TB	Aqua				A & DHC
E. Schroeder	Bottle(73)	TB	Aqua				
(Reference to Ed Schroeder, bottler)	1880 MC						
Geo. Schroeder	Bottle	HB, AC	Aqua		Star		F.H.
(Geo.Schroeder opened new plant)	NBG, 4/1891						
Spannagel Soda & Mineral Water Co.	Bottle(73)	HB	Aqua		S		
(Spannagel Soda & Mineral Water Co. succeeded J.N.Spannagel & Son)	NBG, 6/1891						
EDWARDSVILLE (Madison Co.)							
(Bottling firm of Henry D. Harles established by Franklin Harles in 1871)	NBG, 12/1901						
H. D. Harles	Bottle(73)	HB	Aqua		H		
Franklin Harles	Bottle	HB	Aqua				A & DHC
(Frank Harles, bottler)	1880 MC						
EFFINGHAM (Effingham Co.)							
Effingham Bottling Works	Bottle	ABM-P	Clear	7 oz.	A.V. Conts. 7 fl.oz.		
(Reference to Effingham Bottling Works)	NBG, 4/1930						
Jos. Underreiner Bottler	Bottle	HB	Aqua		JU		
Jos. Underreiner Bottler	Bottle(79)	AC	Aqua		U		Root
ELDORADO (Saline Co.)							
N. Schwartz Bottling Works	Bottle	Crown	Aqua	7 oz.			
(Wink Bott. and Ice Cream Co.)	NBG, 5/1925						
ELGIN (Kane Co.)							
Leo A. Danner Bottling Works	Bottle	HB	Aqua		16		
(Wm. Dettmer, soda & mineral water bottler)	1885 CD						
(Reference to W.M.Dittwer, bottler)	1880 MC						
Elgin Bottling Works	Bottle	HB	Clear				
(Elgin Ice & Beverage Co.)	NBG, 7/1925						
(Samuel Larrew, soda water bottler)	1885 CD						
(Chris Schock purchased business from S. Larrew)	NBG, 11/1899						
Zonian Springs	Fitch 1927						

Embossed Company &/or Bottler Name. Comments: (Owner's name, date(s) of operation, ownership changes, etc.)	Source of Information (Plate Illus.)	Bottle Style	Bottle Color	Vol.	Major Bottom Marks	Trade Mark	Bottle Manuf.
FAIRBURY (Livingston Co.)							
Fairbury Bottling Works/E.T.Clemens	Bottle(79)	ABM	Aqua	8 oz.	F.B.W.		
(J. T. Clemens, bottler)	NBG, 9/1886						
FAIRFIELD (Wayne Co.)							
Fairfield Bottling Works	Bottle	AC			McL., L.R.R.		
FARMINGTON (Fulton Co.)							
(Morrison Union Bott.Works)	NBG, 1/1905						
FLORA (Clay Co.)							
Flora Bottling Works	Bottle(79)	HB, ABM	Aqua				
(C.A.Bruner, proprietor of Flora Bottling Works)	NBG, 4/1915						
Chas. Sebastian Bottler	Bottle	AC	Aqua				
FREEBURG (St. Clair Co.)							
C. H. Heiligenstein	Bottle	HB	Aqua				I.G.Co.
Freeburg Bottling Works	Bottle(79)	ABM	Aqua	6½oz.	J.J.K.		
FREEPORT (Stephenson Co.)							
(Baier and Ohlendorf, bottlers)	NBG, 7/1891						
(Galloway & Lyman succeeded Johns and Lyman	NBG, 5/1896						
(Galloway and Snooks, bottlers)	NBG, 7/1886						
John Schmich	Bottle	AC	Aqua				
(Matt Schmich & Co., bottlers)	1889-90 CD						
(Jacob O. Young, bottler)	NBG, 3/1915						
FROGTOWN (Clinton Co.)							
H. Burger	Bottle(73)	TB	Aqua				
FULLERSBURG (DuPage Co.) (P.O. estab.2/1859, discont.9/1908)							
C. T. Coe	Bottle	HB	Aqua			Anvil	
H. Schramm	Bottle	HB	Aqua		H.S.		NBBG Co.
Henry Schramm	Bottle	HB	Aqua				
Henry Schramm	Bottle(25)	AC	Aqua			Anvil	NBBG Co.
GALENA (Jo Daviess Co.)							
Blum's Bottling Works	Bottle(79)	ABM	Green	7 oz.	BLUM'S		Root
City Bottling Works (F.Blum)	Bottle(79)	AC	Clear		F.BLUM		
City Bottling Works (F.Blum)	Bottle	ABM	Green	8 oz.	F.BLUM		Root
Clem Bussan	Bottle(79)	ABM	Green		ROOT		Root
Clem Bussan	Bottle	ABM	Aqua				W.F.&S. MIL.
Martin Bussan	Bottle	ABM	Green	7 oz.			
(C. E. Kleis, soda water bottler)	1860 MC						
Ch. Plath	Bottle	HB	Aqua				
GALESBURG (Knox Co.)							
C. & S.	Bottle	HB	Aqua		35		
(W.C.Clemens & J.J.Ward; John P. Hickey,Jr.; H.W.McCune: listed under "Bottling Works")	1892-93 CD						
(A. Davis, bottler)	NBG, 8/1896						
Jas. (or James) E. Eaton	Bottle	TB	Aqua	quart			FA & Co.
(Reference to J.E.Eaton, soda water bottler)	1860 MC						
George Edwards	Bottle	TB	Aqua				
(Reference to Camp? & Edwards, soda water bottlers)	1870 MC						
I. R. Edwards	Bottle	HB	Aqua		E		IG Co.
(J. R. Edwards, bottler)	NBG, 10/1886						
E.B.W.	Bottle	HB	Aqua				
Enterprise Bottling Works	Bottler	AC	Clear			EBW	
(Name changed to Galesburg Mfg.Co.)	NBG, 3/1901						
Galesburg	Bottle	ABM-P	Green		PAT. APPLIED FOR		Root

Embossed Company &/or Bottler Name. Comments: (Owner's name, date(s) of operation, ownership changes, etc.)	Source of Information (Plate Illus.)	Bottle Style	Bottle Color	Vol.	Major Bottom Marks	Trade Mark	Bottle Manuf.
GALESBURG (Knox Co.) Cont.							
Galesburg Bottling Works/A. Davis Prop'r.	Bottle	HB	Aqua		D		
Galesburg Steam Bottling Works	Bottle	AC	Aqua		C		
Galesburg Steam Bottling Works	Bottle(79)	AC	Aqua		H		
Galesburg Steam Bottling Works	Bottle	ABM	Aqua	7½oz.	C		
Hoffman Beverage Co.	Bottle(80)	ABM	Clear or Aqua	7½oz.	H		
Hoffman Beverage Co.	Bottle	ABM-P	Amber	7½oz.	GALESBURG ILL.		I
(Reference to D.B.Inness: soda & beer)	1880 MC						
J. R. Inness & Co.	Bottle	HB	Aqua		J.R.I. & Co.		
Sinclair Bros. Bottlers	Bottle(80)	AC	Clear				
Sinclair Bros. Bottlers	Bottle	ABM	Aqua	7½oz.	S		
Sinclair	Bottle(80)	ABM	Clear or Aqua	6½oz. or 7 oz.	S		
GALVA (Henry Co.)							
Galva Bottling Works	Bottle	HB	Aqua				
GENEVA (Kane Co.)							
Juniper Spring Bottling Company	Bottle	ABM	Aqua	7 oz.	Juniper		
GIBSON [GIBSON CITY] (Ford Co.)							
A.B.Fidler's Bottling Wks.	Bottle	AC	Aqua		16		
GIBSON CITY (Ford Co.)							
L. F. Richards Bottling Works	Bottle	AC	Aqua				
GILLESPIE & BENLD (Macoupin Co.)							
(Gillespie Bottling Works: John Hoffes sold to Kam Newman)	NBG, 8/1925						
L. Wilken Bottling Co.	Bottle(73)	HB	Aqua				
GIRARD (Macoupin Co.)							
Girard Bottling Works/J.F.Birzell	Bottle(73)	HB	Aqua				I G Co.
Girard Bottling Works/J.H.Cramer	Bottle(73)	HB	Aqua & Clear		16		
GLEN ELLYN (Cook Co.)							
(Apollo Natural Mineral Water Co.)	NBG, 5/1896						
GOLDEN (Adams Co.)							
Golden Soda Water Works	NBG, 10/1891						
GRAFTON (Jersey Co.)							
(Schafer & Bopp Bott.Works: Schafer sold his interest to Bopp & H. Gerdesmeier)	NBG, 3/1915						
GRANITE CITY (Madison Co.)							
F. Wagner	Bottle	AC	Aqua				I G Co.
(Wagner Bottling Works still in business)	NBG, 7/1930						
GRANTFORK (Madison Co.)							
Diamond Spring	Fitch 1927						
GRASS LAKE (Lake Co.)							
(Robert Selter successor to Selter and Watier, bottlers)	NBG, 4/1901						
GRAYVILLE (White Co.)							
Grayville Bottling Works/A.S.Toops, Prop.	Bottle(73)	HB	Aqua				NBBGCo.
(A.S.Toops successor to E.S.Biddle)	NBG, 10/1901						
GREENUP (Cumberland Co.)							
Greenup Mineral Water Co.	Bottle	HB	Aqua		X		
Cumberland Mineral Springs	Fitch 1927						
GREENVILLE (Bond Co.)							
Greenville Bottling Works	Bottle(80)	ABM	Clear	6½oz.	G.B.W.		Root
(Sodas and distilled drinking water)	NBG, 6/1925						
GRIDLEY (McLean Co.)							
Gridley Bottling Works	Bottle	AC	Aqua				

Embossed Company &/or Bottler Name. Comments: (Owner's name, date(s) of operation, ownership changes, etc.)	Source of Information (Plate Illus.)	Bottle Style	Bottle Color	Vol.	Major Bottom Marks	Trade Mark	Bottle Manuf.
HARRISBURG (Saline Co.)							
Harrisburg Bottling Works	Bottle(73)	HB	Clear		N		I G Co.
Wolf's Beverages	Daily Register, Harrisburg; 4/14/53						
(City?: Ross Mineral Springs, Saline Co.)	Fitch 1927						
HAVANA (Mason Co.)							
Dehm & Co.	Bottle(73)	HB	Aqua				A B Co.
L. Dehm	Bottle(73)	HB	Aqua				
Havana Bottling Works	Bottle(80)	ABM	Aqua	7 oz.	7 oz. CAPACITY		
(Christ. Heimseth, soda water bottler)	1880 MC						
HENRY (Marshall Co.)							
T. B.	Bottle	AC	Clear	quart		Horseshoe	
Henry Bottling Works	Bottle(80)	ABM	Aqua	7 oz.	HENRY		
HERRIN (Williamson Co.)							
Herrin Bottling Co.	Bottle(80)	ABM	Aqua				
Herrin Bottling Co.	Bottle	ABM	Aqua	6½oz.	H.B.CO.		
HIDALGO (Jasper Co.)							
Hidalgo Bottling Works/Dr.A.J.J. Austin, Prop.	Bottle	ABM	Aqua				
HIGHLAND (Madison Co.)							
J.B.	Bottle	HB	Aqua				
L & N	Bottle(73)	HB	Aqua				
J. Amsler	Bottle	TB	Aqua				A & DHC
Highland Bottling Works	Bottle(30)	ABM-P	Clear	7 oz.	7 oz.		I
Mueller & Beck	Bottle(73)	TB	Aqua				
Mueller & Beck	Bottle	HB	Aqua				A & DHC
(Reference to Mueller & Beck, bottlers)	1860 MC						
(Charles Schott Soda Factory)	1880 MC						
Jos. Wick	Bottle	HB, AC	Aqua				
HILLSBORO (Montgomery Co.)							
Hillsboro Bottling Works/Olroyd	Bottle(80)	ABM	Clear or Aqua	7½oz.			Root
Hillsboro Mineral Spring Co.	Bottle(73)	HB	Aqua				
Smith Bros.	Bottle(73)	HB	Aqua				I G Co.
HINSDALE (DuPage Co.)							
(Hinsdale Bottling Co.)	NBG, 11/1899						
Hinsdale Bottling Co.	Bottle(73)	HB	Aqua				
HULL (Pike Co.)							
Hull Bottling Works/Wachandler	Bottle	ABM	Clear		H		
ILLINOIS CITY (Rock Island Co.)							
Illinois City Artesian Well	Fitch 1927						
ILLIOPOLIS (Sangamon Co.)							
Kendall Bottling Works	Bottle(73)	HB	Aqua		V.		
(Kendall Bros. Bottling Works)	NBG, 1/1892						
JACKSONVILLE (Morgan Co.)							
P. Braun & Co.	Bottle	HB	Aqua				
Braun & Killan	Bottle	TB	Aqua				W. McC. & Co.
Doolin Bros. Bottling Works	Bottle(73)	HB	Aqua		20		
E.H.D./Doolin	Bottle	ABM-P	Aqua	6½oz.	EHD		
E. H. Doolin Bottling Works	Bottle(74)	HB	Clear		D		
E. H. Doolin Bottling Works	Bottle	ABM	Aqua	8 oz.	D		
(Don. C. Artz purchased Doolin Bottling Works, still in operation, from Mrs. Doolin)	NBG, 5/1930						
Gravel Springs	Bottle(80)	ABM	Clear	7 oz.			
Gravel Springs Co.	Bottle	AC	Clear	quart			
Gravel Springs Co.	Bottle	ABM	Amber	1 pt., 8 oz.			
Gravel Springs Co.	Bottle	ABM	Aqua	7 oz.	G.S.		I

Embossed Company &/or Bottler Name. Comments: (Owner's name, date(s) of operation, ownership changes, etc.)	Source of Information (Plate Illus.)	Bottle Style	Bottle Color	Vol.	Major Bottom Marks	Trade Mark	Bottle Manuf.
JACKSONVILLE (Morgan Co.) Cont.							
Gravel Springs Company	Bottle(27)	AC	Amber	8 oz.	GRAVEL SPRINGS COMPANY, JACK-SONVILLE, ILL.		
Gravel Springs/A Pure Natural Spring Water	Bottle(24)	TB	Aqua	½ gal.	ROOT		Root
Gravel Springs Products	Bottle	ABM-P	Clear	7 oz.	G.S.		
Gravel Springs/Seligman Bros.	Bottle(74)	HB	Clear				I G Co.
Gravel Springs/Seligman Bros.	Bottle(80)	AC	Clear				
Gravel Springs Beverages/The Very Best	Bottle(27)	AC	Amber	4 & 6 oz.	ROOT		Root
Jacksonville Bottling Works	Bottle	HB	Aqua				
A. Kershaw	Bottle	TB	Aqua		C & I		
(Reference to Albert Kershaw, bottler)	1880 MC						
Ricks & Burns Bottling Works	Bottle(74)	HB	Aqua		20		
G. Schermerhorn	Bottle(74)	HB	Aqua				
(C. Schermerhorn, soda water bottler)	1880 MC						
(C. Schermerhorn & Son)	NBG, 11/1891						
Geo. Strasser Bottling Works	Bottle(74)	HB	Aqua				I G Co.
JERSEYVILLE (Jersey Co.)							
A. F. Mitzel	Bottle(80)	AC	Aqua				
JOHNSTON CITY (Williamson Co.)							
(Johnston City Bottling Co. incorporated)	NBG, 3/1915						
JOLIET (Will Co.)							
T. Binzen	Bottle(74)	HB	Aqua				
(Theodore Binzen, bottler)	1906-7 CD						
J. Hynd	Bottle	AC	Aqua			JH	
(Joliet Bottling Co.)	1906-7 CD						
Joliet Citizens Brewing Co.	Bottle(80)	ABM-P	Aqua	7 oz.			
Paiges Bottling House	Bottle	HB	Aqua				
Auguste Ranft	Bottle	HB	Clear		AR	AR	
Henry Ranft	Bottle	HB	Aqua				
(John Ranft, Bottler)	1891 CD						
Ranft's Bottling House	Bottle(74)	HB	Aqua		R		
G. F. Reimers	Bottle(74)	HB	Aqua		ROOT		Root
G. F. Reimers	Bottle	ABM	Aqua	6½ oz.			Root
G. F. Reimers	Bottle(80)	ABM	Aqua				
C. F. Reimers	Bottle	HB	Aqua				
Reimers & Votik	Bottle	TB	Amber		REIMERS & VOTIK, JOLIET ILL.		
(G. F. Scheidt, bottler)	1906-7 CD						
C. J. Smith	Bottle	HB	Aqua				
J. C. Smith	Bottle	HB	Clear & Aqua		S	JCS	
(J. C. Smith, bottler)	1891, 1906-7 CD						
(Reference to J.C.Smith business)	NBG, 6/1896						
Slovenian National Bottling Co.	Bottle(80)	ABM	Aqua		S.N.		
(Union Bottling Works)	1906-7 CD						
Webb & Riley	Bottle	HB	Aqua		W & R	WR	
Webb & Riley	Bottle	AC	Aqua				
Webb & Riley	Bottle(80)	ABM	Aqua		W & R	WR	
KANKAKEE (Kankakee Co.)							
D. H. Kammann	Bottle	HB	Aqua		29		S.B.& G.Co.
KEWANEE (Henry Co.)							
John P. Brady	Bottle	AC	Aqua				Root
Bradys Bottling Works	Bottle	ABM	Aqua	8 oz.			

Embossed Company &/or Bottler Name. Comments: (Owner's name, date(s) of operation, ownership changes, etc.)	Source of Information (Plate Illus.)	Bottle Style	Bottle Color	Vol.	Major Bottom Marks	Trade Mark	Bottle Manuf.
KEWANEE (Henry Co.) Cont.							
D. & V. Bottling Works	Bottle(80)	ABM	Clear or Aqua	7 oz.	D. & V.		
J. Q. Hill & Co.	Bottle(74)	HB	Aqua				A.G.W.
(John Q. Hill: new bottling plant)	NBG, 12/1891						
J. Q. Hill & Son	Bottle	HB	Clear				
Hill & Stancliffe	Bottle(74)	HB	Aqua				AGWL
Kewanee Bottling Works	Bottle	AC	Aqua		K		ABCo.
Kewanee Bottling Works	Bottle	ABM	Aqua	7 oz.			
Kewanee Bottling Works	Bottle	ABM-P	Green	7 oz.	K	Prairie chicken	I
D. H. Spencer	Bottle(74)	HB	Aqua		S		D.O.C.
KINCAID (Christian Co.)							
L. Vecchie & Son Bottling Co.	Bottle	ABM	Clear	6½oz.	L.V.&S.		
LACON (Marshall Co.)							
Chas. Roth Bottling Works	Bottle	HB	Aqua		15		
(Charles Roth purchased bottling works of F. Berkman)	NBG, 11/1891						
LAKE VIEW (Cook Co.)							
C. Hoffmann & Co.	Bottle	HB	Aqua		H		
A. Fredericks	Bottle(74)	HB	Aqua		15	AF	SABCo
LA SALLE (La Salle Co.)							
(Buchmiller & Maher, bottlers)	1898-99 CD						
Ben Doll	Bottle	HB	Aqua		D		
(Eva Doll, bottler)	1898-99 CD						
(Reference to Ben Doll, bottler)	NBG, 10/1891						
Enterprise Bottling Works	Bottle	HB	Aqua		15		
Enterprise Bottling Works	Bottle	ABM-P	Clear	7 oz.			I
(Alex Finkler, soda water bottler)	1870-1880 MC						
Indian Springs	Fitch 1927						
LAWRENCEVILLE (Lawrence Co.)							
Lawrenceville Bottling Works/"Junior"	Bottle(80)	ABM	Aqua	6½oz.	L		NBBG Co.
Lawrenceville Ice & Cold Storage Co.	Bottle	ABM	Aqua		L		
LEBANON (St. Clair Co.)							
Charles Reuther	Bottle(74)	TB	Aqua				A&DHC
M. Rithmann	Bottle(80)	ABM	Clear	6½oz.			Root
LEMONT (Cook Co.)							
J. G. Bolton	Bottle	TB	Cobalt				Wm.McCully & Co.
J. G. Bolton	Bottle	HB	Cobalt				A & DHC
Wold Brothers Factory	Bottle	HB	Aqua				I.G.Co.
(Wold brothers proprietors of Lemont Ice & Bottling Co.)	NBG, 4/1896						
Edward W. Wold	Bottle(80)	ABM	Aqua	8 oz.	8 oz.		
Edward W. Wold/Highball	Bottle	ABM	Aqua	8 oz.	8 oz.		
E. W. Wold/Quality Beverages	Bottle	ABM-P	Clear	7½oz.	EWW		Root
LEWISTOWN (Fulton Co.)							
Depler Mineral Springs	Bottle	AC	Aqua				
(Depler Bottling Works moved into Lewistown)	NBG, 12/1909						
(Machinery & equipment of Lewistown Bottling Works purchased by S.M. Calkins, Prop., Canton Bottling Works)	NBG, 10/1910						
LIBERTYVILLE (Lake Co.)							
Abana Mineral Springs	Fitch 1927						
Purix Spring	Fitch 1927						
LINCOLN (Logan Co.)							
(One of the best and most complete bottling plants: Herman Berg, Owner)	NBG, 11/1905						
Enterprise Bottling Works/Davis & Co., Props.	Bottle(74, 80)	HB, AC	Aqua				Root

Embossed Company &/or Bottler Name. Comments: (Owner's name, date(s) of operation, ownership changes, etc.)	Source of Information (Plate Illus.)	Bottle Style	Bottle Color	Vol.	Major Bottom Marks	Trade Mark	Bottle Manuf.
LINCOLN (Logan Co.) Cont.							
Enterprise Bottling Works/Howard Vaughn, Prop.	Bottle	HB, AC	Aqua				Root
Lincoln Bottling Works	Bottle(74)	HB	Clear				I G Co.
Lincoln Bottling Works	Bottle	AC	Aqua		H.J.M.		
Lincoln Bottling Works	Bottle	ABM	Aqua		LINCOLN, ILL.		
(H. J. May, bottler)	1898 CD						
Howard Vaughn	Bottle(80)	Semi-ABM	Green		ROOT		Root
LITCHFIELD (Montgomery Co.)							
Wm. Gerhardt	Bottle	HB	Aqua				A & DHC M.G.Co.
John Kaburick Bottler	Bottle(74)	HB	Aqua		J.K.		
Litchfield Bottling Works	Bottle(81)	ABM	Aqua		L.B.W.		
Litchfield Bottling Works	Bottle	HB	Aqua		H & T		
(John A. Leonard sold bottling works to Hauser & Tolle)	NBG, 2/1892						
J. Neuber	Bottle	TB	Aqua				A & DHC
Gus. Reese Bottler	Bottle(74)	HB	Aqua				
C. A. Tolle	Bottle	TB	Aqua		CAT		AB
C. A. Tolle	Bottle(74)	HB	Aqua		C.A.T.	CAT	Root
Tolle Bottling Works	Bottle(75)	HB	Aqua				
Tolle Bottling Works	Bottle(75)	HB	Clear		T	T	
C. A. Tolle Bottling Works	Bottle(75)	HB	Aqua		C.A.T.		
F. Weber	Bottle	HB	Aqua				A & DHC
(Reference to Fred Weber Soda Factory)	1880 MC						
MACOMB (McDonough Co.)							
Elwell's Bottling Works	Bottle	ABM	Aqua	6½oz.	E		
Macomb Bottling Works	Bottle(81)	ABM	Aqua	6½oz.	A		
(George & Bryan dissolved the bottling firm; operated now as the Macomb Bottling Works)	NBG, 12/1884						
(Miner & Smith establish new plant)	NBG, 6/1896						
Nu-Wā Beverage Co.	Bottle	ABM-P	Green	6½oz.	NU-WA		
Pappy's Sodas (Coca Cola Bottling Co.)	Bottle(30)	ABM-P	Green	6½oz.	MACOMB ILLINOIS CANTON		
Pure Pop Co./H.M.Elwell, Prop'r.	Bottle(81)	AC	Aqua		E		
Pure Pop Co./H.M.Elwell, Prop'r.	Bottle	ABM	Aqua	8 oz.	E		
Pure Pop Co./H.M.Elwell, Prop'r./ Elwell's $1000.00 Ginger Ale	Bottle	AC	Aqua	quart	E		
MADISON (Madison Co.)							
Gay-Ola Bottling Co.	Bottle(30)	ABM-P	Aqua	6½oz.	Gay-Ola		
MAQUON (Knox Co.)							
Aqua Vitae Spring	Fitch 1927						
MARINE (Madison Co.)							
V. Bernard	Bottle	HB	Aqua				
Jno. A. Gross	Bottle(75)	HB	Aqua				
MARION (Williamson Co.)							
Coal Belt Bottling Co.	Bottle	ABM-P	Clear	7 oz.	CB CAP.7 oz.		
Marion Bottling Co./Crisp's	Bottle	ABM	Clear	12 oz.	MARION BOTTLING CO. MARION, ILL. 12 FL.OZ.		I
MARSHALL (Clark Co.)							
J. B. Bash	Bottle	ABM-P	Aqua	6 oz.	JB		Root
MASCOUTAH (St. Clair Co.)							
N. Bassler & Co.	Bottle	TB	Aqua				
F. Herold	Bottle(75)	TB	Aqua				

Embossed Company &/or Bottler Name. Comments: (Owner's name, date(s) of operation, ownership changes, etc.)	Source of Information (Plate Illus.)	Bottle Style	Bottle Color	Vol.	Major Bottom Marks	Trade Mark	Bottle Manuf.
MASCOUTAH (St. Clair Co.) Cont.							
Philip Pfeifer, Jr.	Bottle	ABM	Green	$7\frac{1}{2}$oz.	P		
Pfeifer Bottling Co.	Bottle	ABM	Clear	$7\frac{1}{2}$oz.	P		I
MATTOON (Coles Co.)							
Union Bottling Works	Bottle(75)	HB	Aqua				NBBG Co.
Union Bottling Works	Bottle	AC	Aqua				
(Reference to J.H.Wachtler bottling plant)	NBG, 6/1896						
MENDOTA (La Salle Co.)							
(Charles Gephard succeeds George E. Blanchard)	NBG, 12/1901						
Mendota Bottling Co.	Bottle(81)	ABM	Aqua	7 oz.	M		
(Mendota Bottling Works owned and operated by the Hebel brothers)	NBG, 11/1925						
(Mollen & Co., soda water bottlers)	1870 MC						
E. Wohlers	Bottle	HB	Clear				
(Reference to E. Wohlers, bottler)	NBG, 8/1886						
MEREDOSIA (Morgan Co.)							
Hilda's Soda	Bottle	AC	Aqua		H		
METAMORA (Woodford Co.)							
(G. E. Eichorn went out of bottling business)	NBG, 8/1891						
METROPOLIS (Massac Co.)							
(F.G.Bergdoll, bottler, moved to Paducah, Ky.)	NBG, 4/1896						
Kreuter Bros./Quality Beverages	Bottle	ABM-P	Clear	7 oz.	CONTS.7 FL. OZS.ELDORADO		N
(Metropolis Bottling Works established in 1908)	NBG, 1/1910						
MILFORD (Iroquois Co.)							
(Reference to Charles Allison, bottler)	NBG, 10/1891						
MILLEDGEVILLE (Carroll Co.)							
(Milledgeville Bottling Co.: owned by Allen Polhill)	Jour.Ill.State Hist.Soc.1964. Vol.57, No.3						
MILLSTADT (St. Clair Co.)							
(Baer Bottling Works)	NBG, 5/1930						
MILTON (Pike Co.)							
Perry Springs	Fitch 1927						
MINONK (Woodford Co.)							
(N.L.Davidson sold bottling works to R.B.Witte)	NBG, 4/1930						
MOLINE (Rock Island Co.)							
F. C. H.	Bottle	HB	Aqua		H		
(F. C. Hoffman, new bottler)	NBG, 5/1891						
(Moline Bottling Works)	NBG, 9/1891						
(Moline Bottling Works, Jos. Scherrer, Prop)	1912 CD						
(J. H. Mueller, bottler)	NBG, 6/1891						
Jos. Scherrer	Bottle(75)	HB	Aqua				
MOMENCE (Kankakee Co.)							
(Ardon Sherwood, bottler)	NBG, 12/1925						
(Momence Bottling Works; Sherwood and Son)	NBG, 3/1930						
MONMOUTH (Warren Co.)							
E.B.W.	Bottle	HB	Clear				
(Herman W. Karstedt, bottler)	1880 MC						
Maple City Bottling Works	Bottle	ABM	Aqua or Clear	7 oz.	M		Root I

Embossed Company &/or Bottler Name. Comments: (Owner's name, date(s) of operation, ownership changes, etc.)	Source of Information (Plate Illus.)	Bottle Style	Bottle Color	Vol.	Major Bottom Marks	Trade Mark	Bottle Manuf.
MONTGOMERY (Kane Co.)							
Aurora Lithia Springs	Fitch 1927						
Montgomery Magnesia Spring Water Co.	Bottle	HB	Aqua		29		
MORRIS (Grundy Co.)							
Tony Anderson	Bottle	ABM	Aqua	6½oz.			
James Hansen	Bottle	TB	Aqua		A B Co		A B Co
James Hansen	Bottle	AC	Aqua		A B Co	JH	A B Co
MT. CARMEL (Wabash Co.)							
Mt. Carmel Bottling Co.	Bottle(70)	ABM	Aqua		Star		Root
Mt. Carmel Ice & Bottling Co.	Bottle(75)	HB	Aqua				NBBG Co.
Mt. Carmel Ice & Bottling Co.	Bottle(81)	AC	Aqua		Star		
Wolfs/Mt. Carmel Bottling Co.	Bottle(70)	ABM-P	Clear	6½oz.	MT.CARMEL ILL.		I
Wolfs Soda (Coca Cola Bottling Co.)	Bottle(70)	ABM-P	Green	6 oz.	MT.CARMEL ILL.		
MT. OLIVE (Macoupin Co.)							
Mt. Olive Bottling Works	Bottle(81)	ABM	Aqua	7½oz.			
MT. STERLING (Brown Co.)							
(Mount Sterling New Life Mineral Water Co.)	NBG, 6/1910						
MT. VERNON (Jefferson Co.)							
Chero-Cola	Bottle(29)	ABM	Aqua	7 oz.	CHERO-COLA		
Green Lawn Springs	Fitch 1927						
Mount Vernon Steam Bottling Works	Bottle(75)	HB	Aqua				AB
(Louis Niemann, bottler)	NBG, 4/1891						
(Thomas Manion successor to L. Niemann)	NBG, 2/1900						
Perfection Soda Water (Coca Cola Bott. Co.)	Bottle(30)	ABM	Green	6 oz.	MT.VERNON ILL.		
MURPHYSBORO (Jackson Co.)							
(Ackerman & Steinle, bottlers, bought out Ed Hayes in Feb., 1891)	NBG, 1/1892						
(Reference to Edward Hayes, mineral water bottler)	1880 MC						
Murphysboro Bottling Co.	Bottle(81)	ABM	Aqua	7 oz.			
(Murphysboro Bottling Co.)	1922 CD						
Simon & Co.	Bottle(75)	HB	Aqua		S		NBBG Co.
Simon & Co.	Bottle(81)	AC	Aqua		S		
(Simon & Co., bottler)	1922 CD						
Simon & Co./Simon Pure Soda Water	Bottle(81)	ABM	Aqua		S		
NASHVILLE (Washington Co.)							
American Carlsbad Springs	Fitch 1927						
Lorenz Bottling Co./Since 1881	Bottle	ABM-P	Clear	7 oz.	NASHVILLE ILL.MIN. CONTS.7 FL.OZS.		
NEW ATHENS (St. Clair Co.)							
R. M. Schlesinger	Bottle	ABM	Clear		R.S.		
Schlesinger Bottling Works	Bottle	ABM	Aqua				
NEW BURNSIDE (Johnson Co.)							
H. L. Hester Bottling Co.	Bottle(81)	AC	Aqua				R G Co.
NEWTON (Jasper Co.)							
Newton Bottling Works	Bottle	HB, AC	Aqua		NEWTON in diamond		
(Newton Bottling Works, owned by W.A.McKinnon & Jos.Long, destroyed by fire)	NBG, 11/1909						
NILES CENTER (Cook Co.)							
(P.O. reestab. 4/1873, changed to Skokie 1931)							
Gabriel & Sigwalt	Bottle	HB	Aqua				A.G.W.L.

Embossed Company &/or Bottler Name. Comments: (Owner's name, date(s) of operation, ownership changes, etc.)	Source of Information (Plate Illus.)	Bottle Style	Bottle Color	Vol.	Major Bottom Marks	Trade Mark	Bottle Manuf.
NORTH CHICAGO (Lake Co.)							
(Frank Opeka, proprietor of Star Bottling Works, sold plant to H. Lindeman and C. & E. Staronawicz)	NBG, 3/1920						
O'FALLON (St. Clair Co.)							
B. Bischoff	Bottle(75)	TB	Aqua		B.B.		
OKAWVILLE (Washington Co.)							
Okawville Bottling Works	Bottle(81)	AC	Aqua		O		A B Co.
Okawville Bottling Works	Bottle	ABM-P	Clear	7 oz.			N
(Okawville Mineral Springs Co.)	NBG, 8/1891						
OLNEY (Richland Co.)							
Wm. Bower	Bottle(13)	TB	Aqua		GRAVITATING STOPPER MADE BY JOHN MATHEWS PAT. OCT. 11, 1864 NEW YORK		
Combs Bros. Bottling Wks.	Bottle(81)	AC	Aqua				
Olney Bottling Works	Bottle	HB	Aqua				
(Olney Bottling Works, John Carrol, Prop.)	1903 CD						
Olney Bottling Wks./B.F.Combs	Bottle	ABM	Aqua		O		
Soda Water (Coca Cola Bottling Co.)	Bottle	ABM	Aqua	6½oz.	COCA-COLA		
OREGON (Ogle Co.)							
(Henry Kennedy purchased the Rock River Bottling Works business)	NBG, 4/1896						
(R.W.Hoyt & Co. starting business)	NBG, 7/1901						
OTTAWA (La Salle Co.)							
(Crotty Bros., bottlers)	1869-70 CD						
(Thomas Crotty, mineral water bottler)	1870 MC						
Crow Beverage Co.	Bottle	ABM	Aqua				
(John Devany installed new Mathews plant)	NBG, 7/1896						
A. Finkler	Bottle(75)	HB	Aqua/ green				
(Alex Finkler moved from Ottawa to La Salle to continue bottling)	NBG, 11/1884						
(John J. O'Kane rebuilding burned factory)	NBG, 1/1911						
Ottawa Bottling Works/Carse & Co.	Bottle(75)	HB	Aqua		20		
Sanicula Mineral Springs	Bottle(75)	HB	Aqua		SANICULA SPRINGS		
Sanicula Mineral Springs	Bottle	ABM	Clear				
W. J. Sinon & Co.	Bottle(75)	HB	Aqua		SANICULA SPRINGS		SAB Co.
PANA (Christian Co.)							
F. J. Weber Bottler	Bottle	AC	Aqua		W		
F. J. Weber Bottler	Bottle	ABM	Aqua	6½oz.	W		A & DHC
F. Weber & Bro.	Bottle	HB					
(Reference to Jacob Weber)	1880 MC						
(Reference to Weber bottling works)	NBG, 10/1901						
PARIS (Edgar Co.)							
(E.A.Patty and Lansaw sold their ice and bottling business to Lawrence Buckler)	NBG, 10/1909						
PAXTON (Ford Co.)							
Victor Hampton Bottling Works	Bottle(75)	HB	Aqua				
Paxton Bottling Co.	Bottle	ABM	Clear	7 oz.	P		
S & G Bottling Co.	Bottle(75)	HB	Aqua				

Embossed Company &/or Bottler Name. Comments: (Owner's name, date(s) of operation, ownership changes, etc.)	Source of Information (Plate Illus.)	Bottle Style	Bottle Color	Vol.	Major Bottom Marks	Trade Mark	Bottle Manuf.
PEKIN (Tazewell Co.)							
Nox-Age Bottling Works	Bottle(81)	ABM	Clear	6½oz.	F & Z		
(D.W.Frederick and W.L.Zimmerman organized the Nox-Age Bottling Works)	NBG, 7/1925						
(Pekin Bottling Works, G.J.Zerwekh Sons Co., Proprs.)	1898 CD						
Pekin Bottling Works/G.J.Z.	Bottle(75)	HB	Aqua		P in triangle	P in triangle TRADE MARK	
Pekin Bottling Works/G.J.Z.	Bottle	AC	Aqua				
Pekin Bottling Works	Bottle(81)	ABM	Aqua	7½oz.	P		
Peek-In/Pekin Bottling Works	Bottle(29)	ABM-P	Clear	7 oz.	CAPACITY 7 OUNCES		I
Red Rock Bottle Co.	Bottle	ABM-P	Green		CAP.7 FL. OZ.		
Tazewell Bottling Works	Bottle	ABM	Clear	6½oz.	T		
(G.Zerwekh & Son purchased bottling works of C.M.Huisken & Co.:	NBG, 11/1891						
Pekin Bottling Works)	NBG, 2/1892						
PEORIA (Peoria Co.)							
Frank Burre/Quality Beverages	Bottle	ABM-P	Clear	7 oz.	FRANK BURRE PEORIA ILL.		
Cairn & Hickey	Bottle	TB	Green		(Pontil mark)		
Central City Bottling Co.	Bottle(76)	HB	Aqua		I.G.Co.		I G Co.
The City Bottling Co.	Bottle	HB, AC	Aqua				
Clark's Peoria Spring Soda	Bottle(76)	HB	Aqua		C		
A. Davis Bottling Works	Bottle(81)	AC	Aqua		AD	AD	NBBG Co.
A. Davis/Peoria Mineral Springs	Bottle(76)	HB	Aqua		AD	AD	
A. Davis/Peoria Bottling Works	Bottle(76)	HB	Clear		AD	AD	
(Wm. Dory purchased interest of Davis Bottling Co.; now named Advance Bottling Company)	NBG, 1/1910						
Derges Bottling Co, /Squeeze	Bottle	ABM-P	Clear	7 oz.	DERGES BOT.CO. PEORIA ILL.	SQUEEZE	
J.E.E.	Bottle	HB	Aqua		E		
J.E.Eaton & Co.	Bottle	HB	Aqua		E		
(Reference to J.E.Eaton, bottler)	1880 MC						
H. Fellrath	Bottle(76)	HB	Aqua		F		
(Reference to H. Fellrath, bottler)	1880 MC						
R. Hickey	Bottle	HB	Aqua		H		FA & Co.
(Reference to R.E.Hickey, bottler)	1860 MC						
Howland & Co.	Bottle	HB	Clear				
Jobethco Brand - Distributors	Bottle (paper label)	ABM	Black	12 oz.			
Kirwan & Tobin	Bottle	HB	Aqua		K & T	Anchor	
McMaster & Derges/I.B.W.	Bottle(76)	HB	Aqua				SAB Co. W.F.&S., Mil.
McMaster & Derges/I.B.W.	Bottle	AC	Aqua				
McMaster & Derges/I.B.W.	Bottle	ABM	Aqua	7 oz.			
R. L. Pasquay	Bottle	HB	Aqua		R.L.P.		
Peoria Bottling Co.	Bottle	HB	Clear				
(Reference to Peoria Bottling Co.)	NBG, 9/1886						
Peoria Bottling Works/Korn Bros. & South	Bottle	ABM	Green	8 oz.			
Chas. Singer	Bottle	HB	Aqua		C.S.		C & Co. SB & G Co.
(Reference to Charles Singer, bottler)	1880 MC						
The Singer Bottling Co.	Bottle	HB	Aqua				R G Co.
The Singer Bottling Co.	Bottle	AC	Aqua		S		Root
The Singer Bottling Co.	Bottle	ABM	Aqua	8 oz.	S		

Embossed Company &/or Bottler Name. Comments: (Owner's name, date(s) of operation, ownership changes, etc.)	Source of Information (Plate Illus.)	Bottle Style	Bottle Color	Vol.	Major Bottom Marks	Trade Mark	Bottle Manuf.
PEORIA (Peoria Co.) Cont.							
Singer Bros.	Bottle	HB	Aqua		S.B.		SB&GCo.
Singers	Bottle	ABM	Aqua	6½oz.	S		Root
Union Bottling Works	Bottle	ABM	Clear	7 oz.	U.B.		
Williams & Buchele/Arcana	Bottle	ABM	Clear	6½oz.	A		
PERU (La Salle Co.)							
J. G. Mischke	Bottle	AC	Aqua		M		
Henry Soedler	Bottle	HB	Aqua				D.O.C.
(Reference to Henry Soedler)	NBG, 6/1891						
(Henry Soedler, bottler)	1898-99 CD						
Star Union Products Co.	Bottle	ABM-P	Clear	7½oz.	Star		
PETERSBURG (Menard Co.)							
O. F. Lenz	Bottle	HB	Aqua				
O. Lenz	Bottle	HB	Aqua			OL	NBBG Co.
(Reference to Otto F. Lenz)	NBG, 12/1891						
Lenz's	Bottle	AC	Aqua		L		
Petersburg Bottling Works	Bottle(81)	ABM	Aqua	6½oz.	L		
PINCKNEYVILLE (Perry Co.)							
(Pinckneyville Bottling Works?) Pinckneyville, Ill.	Bottle(29)	ABM-P	Clear	7 oz.	7 FL.OZ.		N
Pinckneyville Bottling Works	Bottle(30)	ABM-P	Clear	6½oz.			
PITTSFIELD (Pike Co.)							
Pittsfield Bottling Works	Bottle	AC, ABM	Aqua	7½oz.	P		
PONTIAC (Livingston Co.)							
(Reference to H.H.Scatterday, bottler)	NBG, 11/1884						
H. H. Scatterday	Bottle(76)	HB	Clear		16		
H. H. Scatterday	Bottle(81)	ABM	Aqua	7 oz. or 8 oz.			
(The Scatterday Co., Loras A. Ellis, Prop.)	1926 CD						
Scatterday's Bottled Beverages	Bottle	ABM	Aqua	7 oz.			
E. T. Yarke Bottling Co.	Bottle	ABM	Aqua	7 oz.	Y		
PRINCETON (Bureau Co.)							
(William H. Lynch, bottler)	NBG, 7/1884						
G. W. Simon & Bro.	Bottle	HB	Aqua		X		D.O.C.
QUINCY (Adams Co.)							
Bergmann's Bott. Wks.	Bottle(81)	ABM	Aqua	6½oz.	B		
Bergman's Bottling Works	Bottle(24)	ABM	Aqua	24 oz.	B		
(Bergman Bottling Works took over the Mills Co. about 1920-1921)	Pers.Comm. Carl Landrum, Quincy. 5/14/70						
Boschulte & Co.	Bottle	TB	Aqua				A & DHC
H. Durholt & Co.	Bottle(76)	HB	Aqua				
Henry Durholt	Bottle	HB	Aqua				
(Durholt started bottling in late 1860s; out of business ca. 1898)	Pers.Comm. Carl Landrum 5/14/70						
(Reference to H. Durholt & Co.)	1880 MC						
(Fisher & Boschulte, bottlers in 1861; Boschulte & Knauf, bottlers in 1869)	Pers.Comm. Carl Landrum 5/14/70						
J. J. Flynn & Co.	Bottle(76)	HB	Aqua		F		
J. J. Flynn & Co.	Bottle(81)	AC/ ABM	Aqua		F		
J. J. Flynn & Co.	Bottle(24)	AC	Aqua	ca.24 oz.	ABCo	JJF	ABCo.
J. J. Flynn & Co.	Bottle	ABM	Amber	ca.24 oz.			
J. J. Flynn & Co.	Bottle(81)	ABM	Aqua	over 5 oz.	F		
J. J. Flynn & Co.	Bottle(30)	ABM-P	Clear & Aqua	6 oz.	F		
(John J. Flynn established bottling business ca. 1880; in the beginning bottled only spruce & root beer)	Pers.Comm. Carl Landrum 5/14/70						

Embossed Company &/or Bottler Name. Comments: (Owner's name, date(s) of operation, ownership changes, etc.)	Source of Information (Plate Illus.)	Bottle Style	Bottle Color	Vol.	Major Bottom Marks	Trade Mark	Bottle Manuf.
QUINCY (Adams Co.) Cont.							
(Reference to J. J. Flynn & Co.)	1880 MC						
(J.J.Flynn added new plant for distilling water; now producing mineral and lithia water)	NBG, 8/1896						
(Grone & Durholt, bottlers)	1860 MC						
(Lampe & Boschulte, soda water bottlers)	1860 MC						
L and B Soda	Bottle	AC	Aqua		Q		
L and M Soda	Bottle(81)	AC	Aqua		Q		
(Fred Lubker and George Montag started business about 1898)	Pers.Comm. Carl Landrum 5/14/70						
Mills Bros. & Co.	Bottle	HB,ABM	Aqua		M		
Mills Bros.	Bottle(82)	AC	Aqua		M		
Mills Bros. Bottling (or Bott.Wks.)	Bottle	ABM	Aqua		M		
(Mills Bros., Wm. & George, starting a new bottling firm)	NBG, 4/1896						
Quincy Bottling Works	Bottle	ABM	Aqua	over 7 oz.	OVER 7 Q OUNCES		
Quincy Bottling Works/Lubker's Big 9	Bottle	ABM	Clear	9 oz.	Q		
Waggener Bros.	Bottle(82)	AC	Aqua		W.BROS. & CO.		
(Waggener Bros. operated a saloon and bottling company only a short time: ca. 1910)	Pers.Comm. Carl Landrum 5/14/70						
RED BUD (Randolph Co.)							
Fred Husemann Bottling Works	Bottle	ABM-P	Clear	7 oz.	G in a square		
Lohrberg & Ratz	Bottle(76)	HB	Aqua		22		IGCo.
RIPLEY (Brown Co.)							
(New Life Mineral Springs Co. to begin bottling operations)	NBG, 4/1910						
RIVER FOREST (Cook Co.)							
Adolph Westphal	Bottle	HB	Aqua				
Westphals	Bottle	ABM	Aqua	1 pt., 8 oz.	ADOLPH WESTPHAL RIVER FOREST		I
ROBINSON (Crawford Co.)							
Robinson Bottling Works	Bottle	AC	Aqua		ROOT		Root
Robinson Bottling Works	Bottle	ABM	Aqua		ROBINSON ILL.		
ROCHELLE (Ogle Co.)							
(E. Gardner, bottler)	NBG, 6/1891						
ROCK FALLS (Whiteside Co.)							
(Reference to James J. Allison, bottler)	NBG, 3/1884						
ROCK ISLAND (Rock Island Co.)							
Blackhawk/Ginger Ale (Probably Carse & Ohlweiler)	Bottle(82)	ABM	Aqua				
Black Hawk Spring	Fitch 1927						
C. & O.	Bottle	HB	Aqua				S.B. & G.Co.
C & O	Bottle	AC	Aqua			C & O	C & Co., Lim.
C & O	Bottle	ABM	Aqua			C & O	
Carse & Ohlweiler	Bottle(76)	TB	Aqua		C & O		
(Carse & Ohlweiler, bottlers)	1880 MC						
(Henry Carse opened bottling plant in Rock Island in 1868; Ohlweiler joined the firm in 1872)	NBG, 7/1884						
A. D. Huesing	Bottle(25, 82)	HB,AC	Aqua		ADH	ADH	C & Co., Lim.

Embossed Company &/or Bottler Name. Comments: (Owner's name, date(s) of operation, ownership changes, etc.)	Source of Information (Plate Illus.)	Bottle Style	Bottle Color	Vol.	Major Bottom Marks	Trade Mark	Bottle Manuf.
ROCK ISLAND (Rock Island Co.) Cont.							
(Albert D. Huesing Bott. Works)	1912 CD						
(Rock Island Artificial Ice Co.)	NBG, 5/1925						
(Rock Island Bottling Works/Julius Junge)	NBG, 11/1891						
(Tri-City Bottling Co.)	NBG, 10/1910						
Woodman Artesian Well	Fitch 1927						
ROCKFORD (Winnebago Co.)							
(Boyer Brothers, bottlers)	NBG, 4/1915						
J. M. Clarke	Bottle	HB, AC	Aqua				
J. M. Clarke	Bottle(82)	Semi-ABM	Aqua		ROOT		Root
J. A. Holland	Bottle(76)	HB	Aqua		C36		
(Bob Lincoln, bottler)	NBG, 8/1884						
(George Lincoln, bottler)	NBG, 3/1884						
(Alfred Lundquist succeeds C.J. Nelson)	NBG, 8/1896						
(Rockford Hygiene Spring Co.)	NBG, 11/1891						
(P. W. Welch, Bottler)	NBG, 7/1901						
Wilson	Bottle	ABM	Aqua	7½oz.	W		
Wilson Bottling Co., Inc.	Bottle	ABM-P	Aqua	7 oz.			I
(Reference to Wilson Bottling Co.)	NBG, 12/1925						
ROODHOUSE (Greene Co.)							
Roodhouse Bottling Works	Bottle	HB	Aqua		F		
Roodhouse Bottling Works	Bottle	AC	Clear		R		
Friedman's Bottling Works	Bottle	AC	Aqua		DSF		
RUSHVILLE (Schuyler Co.)							
Rushville Bottling Works	Bottle(82)	AC	Aqua				
Rushville Bottling Wks.	Bottle(30)	ABM-P	Clear	6½oz.	R.B.W.		I
(City ?: Schuyler County Spring)	Fitch 1927						
SAILOR SPRINGS (Clay Co.)							
Sailor Springs	Fitch 1927						
SARATOGA (Union Co.)							
Hygienic Springs	Fitch 1927						
SAVANNA (Carroll Co.)							
(O.D.Smith moved bottling works to Iowa)	NBG, 11/1910						
(J.H.Gerbach succeeds business purchased by R.B.Myers from Geo.Wicks)	NBG, 3/1915						
SHELBYVILLE (Shelby Co.)							
O. L. Lewis	Bottle	AC	Aqua		O		
SOUTH WILMINGTON (Grundy Co.)							
Anton Fenoglio	Bottle	HB	Aqua				A B Co.
SPARTA (Randolph Co.)							
Sparta Bottling Works	Bottle	ABM	Aqua	6½oz.			I
SPRINGFIELD (Sangamon Co.)							
F. M. Busby	Bottle(76)	HB	Aqua				
Capital Beverage Co.	Bottle	ABM	Aqua	6 oz.	C		
Capital City Bottling Works/Ihlenfelt & Jurgens	Bottle	ABM	Aqua				
Chero Cola Bottling Co./Soda Water	Bottle	ABM-P	Clear & Aqua	6 oz.	PAT'D JUNE 3 1924		I
Chero Cola Bott. Co./Sodawater	Bottle(30)	ABM-P	Green	6 oz.	PATENT PENDING		
Chero Cola	Bottle(30)	ABM-P	Clear	6 oz.	PATENT PENDING		
C.C.Soda/Coca Cola Bottling Co.	Bottle(30)	ABM-P	Green & Clear	6 oz.	SPRINGFIELD ILL.		Root
W. W. Clutter	Bottle(76, 82)	HB, AC Semi-ABM	Aqua		C		Root

Embossed Company &/or Bottler Name. Comments: (Owner's name, date(s) of operation, ownership changes, etc.)	Source of Information (Plate Illus.)	Bottle Style	Bottle Color	Vol.	Major Bottom Marks	Trade Mark	Bottle Manuf.
SPRINGFIELD (Sangamon Co.) Cont.							
W. W. Clutter	Bottle	ABM	Aqua	6½oz.	Star		
Clutter's Quality	Bottle(82)	ABM	Green	10 oz.			
(Walter Clutter, employee at Springfield Mineral Spring Works, injured by an exploding bottle of soda)	NBG, 7/1901						
(T.C.Cole purchased bottling business of B. Meyer)	NBG, 12/1909						
(J. E. Craft purchased the Clutter Bott. Co.)	NBG, 9/1925						
Davis & Marshall	Bottle	HB	Aqua		D & M		NBBGCo.
(L. Davis is successor to Mrs. M.A. Fischer)	NBG, 2/1900						
(L. Davis & Co.;W.J.Grant; John Lauterbach; C.J.Peterson: "Soda & Mineral Water Mnfrs.")	1900-01 CD						
H. W. Dresch	Bottle(82)	ABM	Clear & Aqua	7½oz.			I
Elliott & Cullom	Bottle	HB	Aqua				
(Fehr and Hofferkamp, bottlers)	NBG, 8/1891						
M. A. Fischer	Bottle(76)	HB	Aqua				D.O.C.
(Reference to M.A.Fischer, bottler)	1880 MC						
(Reference to M.A.Fischer)	NBG, 10/1891						
Grant & Co.	Bottle(76)	HB	Aqua		REGISTERED G		I.G.Co.
Grapico Bottling Works	Bottle(82)	ABM-P	Aqua	6 oz.	G		
Grapico Bottling Works/The drink of the Nation	Bottle	ABM-P	Aqua	6½oz.			
Halls Wellhall Water	Bottle	ABM	Aqua	7 oz.			
Wellhall	Bottle(82)	ABM	Aqua	7 oz.			
Johnson & Peterson	Bottle(76, 77)	TB, HB	Aqua				
(Reference to Johnson & Peterson, bottlers)	1880 MC						
(Reference to Johnson & Peterson)	NBG, 4/1891						
John Lauterbach	Bottle(77)	HB	Aqua		L		D.O.C.
Lauterbach & Reisch	Bottle(77, 82)	HB, AC	Aqua		L & R		Root
Lime Cola	Bottle(29)	ABM-P	Green	6 oz.	SPRINGFIELD ILL.		Root
Marshall Bottling Co.	Bottle(77)	HB	Aqua		M.B.Co.		NBBGCo.
Mester Bottling Co.	Bottle	ABM-P	Clear		PAT'D JUNE 3, 1924		N
Mester's Ginger Ale	Bottle	ABM	Green	10 oz.			
Nu-Grape	Bottle	ABM-P	Green	6 oz.	SPRINGFIELD ILL.		Root
Nu Grape Bottling Co.	Bottle	ABM-P	Clear	6½oz.	N.B.C.		
C. J. Peterson	Bottle(77)	HB	Aqua		P		N.B.B.G.Co. I.G.Co.
F. X. Reisch	Bottle(77)	HB, AC	Aqua		R		S.A.B.Co.
Reisch	Bottle	ABM	Aqua		R		
Reisch Brg. Co.	Bottle	ABM	Clear & Green	7 oz.	R		Root
Rule Bros. Bottlers	Bottle	HB	Aqua		R		
Springfield Mineral Springs/L.Davis & Co.	Bottle(76)	HB	Aqua		D		D.O.C.
SPRING VALLEY (Bureau Co.)							
Spring Valley Bottling Works	Bottle(82)	HB, AC	Aqua		D		I.G.Co.
Spring Valley Bottling Works	Bottle	ABM	Aqua	7 oz.	D		
ST. JACOB (Madison Co.)							
St. Jacob Bottling Works/Wm.Noeltner	Bottle	TB	Aqua				
St. Jacob Bottling Works	Bottle	AC	Aqua		FB		
STAUNTON (Macoupin Co.)							
Luker Bros.	Bottle(77)	HB	Aqua		L		Root

Embossed Company &/or Bottler Name. Comments: (Owner's name, date(s) of operation, ownership changes, etc.)	Source of Information (Plate Illus.)	Bottle Style	Bottle Color	Vol.	Major Bottom Marks	Trade Mark	Bottle Manuf.
STAUNTON (Macoupin Co.) Cont.							
Luker Bros.	Bottle	AC	Clear		$7\frac{1}{2}$oz.		
Luker Bros./Distilled Water	Bottle(24)	TB	Aqua	$\frac{1}{2}$gal.	A B Co		A B Co.
R. Stille	Bottle(77)	HB	Aqua				I G Co.
Teufhert Brothers	Bottle	ABM	Aqua	7 oz.	T.B.		
STERLING (Whiteside Co.)							
(W.C.Clemens, bottler)	NBG, 10/1886						
(F.H.Johns, bottler)	NBG, 7/1896						
(Middleton Bottling Works)	NBG, 5/1930						
STREATOR (La Salle Co.)							
American Bottling Works/W. Messersmith	Bottle(82)	ABM	Aqua	8 oz.			
(Moses Ashcroft, mineral water bottler)	1898-99 CD						
(Harry H.Dietman, mineral water bottler)	1898-99 CD						
(Finkler BottlingWorks)	NBG, 8/1891						
Harcharik Bottling Works	Bottle	ABM-P	Aqua	7 oz.	C		I
(Richard Murphy, bottler)	NBG, 4/1891						
(Haney and Dietman succeed R.E. Murphy)	NBG, 5/1896						
J. H. Ridings & Sons	Bottle	HB	Aqua		C		SAB Co.
(Ryan Bottling Co.)	1898-99 CD						
Streator Bottling Works/Morris Ashcroft	Bottle	HB	Aqua		20		
SYCAMORE (DeKalb Co.)							
(B.Rogers succeeds Montgomery and West)	NBG, 4/1915						
TAYLORVILLE (Christian Co.)							
C.C.C.	Bottle(77)	HB	Aqua				
Crystal Carbonating Co./Frank M. Bonheims/LICOR Mfg.Co.	Bottle	ABM	Aqua	$6\frac{1}{2}$oz.			
(R.I.Smith moved his bottling plant from Litchfield to Taylorville)	NBG, 4/1891						
(Star Bottling Works)	1928-29 CD						
TEUTOPOLIS (Effingham Co.)							
Jos. H. Buehnerkempe Bottling Works	Bottle	ABM	Aqua	7 oz.			Root
THORNTON (Cook Co.)							
(Thornton Mineral Springs Co., F.M. Miller, Pres.)	NBG, 8/1925						
TINLEY PARK (Cook Co.)							
Philip Mager	Bottle	HB	Aqua		S.H.		A.G.W.
(F.Schmaedeke sold his bottling works to H.Brodthuhn; will open new factory in Paxton)	NBG, 12/1891						
Tinley Park Bottling Works	Bottle(77)	HB	Aqua				NBBG Co.
Tinley Park Bottling Works/G.Goebel	Bottle(82)	ABM	Aqua	7 oz.	G		I
Fred Wilke	Bottle	HB	Aqua				Root
TOLUCA (Marshall Co.)							
Toluca Bottling Works	Bottle(30)	ABM-P	Clear	7 oz.	T		I
(Toluca Bottling Works operating in 1897)	Pers.Comm., Mrs.F.Carrithers, Toluca:10/21/71						
TRENTON (Clinton Co.)							
(Trenton Bottling Works/Geo.Keck)	NBG, 3/1915						
TROY (Madison Co.)							
Newport Soda Co.	Bottle	ABM-P	Clear	7 oz.			
(Reference to Newport Soda Co.)	NBG, 3/1930						
TUSCOLA (Douglas Co.)							
Nu Icy	Bottle	ABM-P	Clear	$\frac{1}{2}$ pint	TUSCOLA ILL.		Root

Embossed Company &/or Bottler Name. Comments: (Owner's name, date(s) of operation, ownership changes, etc.)	Source of Information (Plate Illus.)	Bottle Style	Bottle Color	Vol.	Major Bottom Marks	Trade Mark	Bottle Manuf.
TUSCOLA (Douglas Co.) Cont.							
Tuscola Bottling Works	Bottle	ABM	Green	$6\frac{1}{2}$ oz.	T.A.		
(Tuscola Bott.Works owned and operated by Frank R.Smith; bottled 15 different soda drinks, bottled 250 cases daily. ca.1925)	Pers.Comm. R.E.Owen, Tuscola:5/7/71						
URBANA (Champaign Co.)							
Dudenhofer Bottling Co.	Bottle(82)	ABM	Green	7 oz.			Root
Twin City Bottling Works	Bottle(82)	AC	Aqua				M.B.& G.Co.
VALMEYER (Monroe Co.)							
Valmeyer Bottling Works	Bottle	ABM	Aqua				
VANDALIA (Fayette Co.)							
(Rene Butel, bottler)	NBG, 1/1885						
(Dean & Henry, bottlers, succeed Fred Smith)	NBG, 7/1896						
A. G. Meyer	Bottle(77)	HB	Aqua & Clear		20		I G Co. NBBG Co.
W. F. Smith	Bottle(77)	HB	Aqua		Bottle		NBBG Co.
Vandalia Bottling Works	Bottle	AC	Aqua		V.B.W.		
V.B.W. (on bottom, only)	Bottle	AC	Aqua & Clear		V.B.W.		
Vandalia Dairy Co./Ice Cream & Bot. Wks./Hudsons	Bottle	ABM-P	Clear	7 oz.	VANDALIA ILL.		I
VENICE (Madison Co.)							
Tri-City Bottling Co.	Bottle	AC	Clear		V		
Wagner & Co.	Bottle	AC	Aqua				I G Co.
VERSAILLES (Brown Co.)							
Versailles Springs	Fitch 1927						
VIRDEN (Macoupin Co.)							
Achilles & Hille	Bottle	HB	Aqua				
Chero Cola Bottling Co./Nehi	Bottle	ABM-P	Clear	9 oz.	DESIGN PAT'D Mar.3, 25		
Frankie & Rich Bottling Works	Bottle	ABM	Aqua		F. & R.		
(Jacob Ring bought out C.E.Hensen)	NBG, 12/1901						
Wm. Griffiths Bottler	Bottle(77)	HB	Aqua		G		
Virden Bottling Works/Wm.Rigby	Bottle(77)	HB	Aqua		V.		N.B.B.G.Co
Virden Bottling Works/Wm.Rigby	Bottle(82)	AC, ABM	Aqua		V.		
Virden Bottling Works	Bottle	ABM	Clear	7 oz.	V.B.W.		
WARREN (Jo Daviess Co.)							
(Miller Bottling Works)	NBG, 7/1925						
WATERLOO (Monroe Co.)							
L. Abegg	Bottle	TB	Aqua				
(Henry Bakey, Soda Factory)	1880 MC						
H. Boeke	Bottle(77)	HB	Aqua		H.B.		I.G.Co.
Charles Boeke	Bottle	Crown	Aqua				
Geo. De Puyt	Bottle	TB	Aqua		Iron pontil		
(Louis Lenhardt bought William Tebbenhoff Soda Works)	NBG, 6/1925						
Henry Schroeder/W.S.F.	Bottle(82)	ABM	Aqua		S		
H. Schuell	Bottle	TB, HB	Aqua				
(Reference to H.S.Schuell)	NBG, 1/1892						
WATSEKA (Iroquois Co.)							
(Consolidated Bottling Works)	NBG, 4/1915						
(Watseka Bottling Works)	NBG, 4/1915						
(Watseka Bottling Works; plant moved from Fairbury. Geo.W.Lace)	NBG, 12/1891						

Embossed Company &/or Bottler Name. Comments: (Owner's name, date(s) of operation, ownership changes, etc.)	Source of Information (Plate Illus.)	Bottle Style	Bottle Color	Vol.	Major Bottom Marks	Trade Mark	Bottle Manuf.
WAUKEGAN (Lake Co.)							
(Durkin Bros. purchased bottling plant of Dolen & Broecker)	NBG, 3/1920						
Glen Flora Spring	Fitch 1927						
Glen Rock Spring Co.	Bottle(77)	HB	Aqua				A.C.W.L.
WEBSTER CITY (County?)							
(Webster City Bottling Works)	NBG, 3/1896						
WESTERN SPRINGS (Cook Co.)							
Alcyone Spring	Fitch 1927						
Keils Deep Well Ginger Ale (Julius M. Keil)	Bottle	ABM	Aqua				
WEST FRANKFORT (Franklin Co.)							
American Bott. Co.	Bottle	ABM-P	Aqua or Clear	6½oz.	W.F.		I
WEST PULLMAN (Cook Co.)							
(P.O. estab. 6/1892, discont. 6/1894)							
Calumet Bottling Works/M.Cohs	Bottle	HB	Aqua		CBW		
WESTVILLE (Vermilion Co.)							
Star Bottling Works	Bottle	TB	Amber		P.J.D.		
Star Bottling Works	Bottle(77)	HB	Aqua		Star		
Star Bottling Works	Bottle	HB	Aqua		P.J.D.		
Star Bottling Works/P.J.Dowiatt	Bottle	HB	Clear		P.J.D.		
Star Bottling Works/P.J.Dowiatt	Bottle	HB	Aqua	quart	P.J.D.		R.G.Co.
Westville Bottling Works/J.D.Gaines, Prop.	Bottle	HB	Aqua		G		
WILLISVILLE (Perry Co.)							
Willisville Bottling Works	Bottle(29)	ABM-P	Clear	6½oz.	B		
WINDSOR (Shelby Co.)							
Windsor Bottling Works	Bottle	ABM	Clear	6½oz.	W.B.W.		
WOOD RIVER (Madison Co.)							
Sweetins Beverages	Bottle	ABM-P	Clear				N
WOODSTOCK (McHenry Co.)							
Peter E. Rose Bottler	Bottle	HB	Aqua				I.G.Co.
WYOMING (Stark Co.)							
Wyoming Bottling Works	Bottle(82)	AC	Aqua				
Wyoming Bottling Works	Bottle(82)	ABM	Aqua	8 oz.			
Wyoming Bottling Works/T.R.	Bottle	HB	Aqua				

Figure 71. Blob-top soda bottles used by Illinois firms (Alton to Carlinville).

Figure 72. Blob-top soda bottles used by Illinois firms (Carlinville to Decatur).

FIGURE 73. Blob-top soda bottles used by Illinois firms (East St. Louis to Jacksonville).

FIGURE 74. Blob-top soda bottles used by Illinois firms (Jacksonville to Litchfield).

FIGURE 75. Blob-top soda bottles used by Illinois firms (Litchfield to Pekin).

FIGURE 76. Blob-top soda bottles used by Illinois firms (Peoria to Springfield).

FIGURE 77. Blob-top soda bottles used by Illinois firms (Springfield to Westville).

Figure 78. Crown soda bottles used by Illinois firms (Altamont to Chicago).

FIGURE 79. Crown soda bottles used by Illinois firms (Chicago to Galesburg).

FIGURE 80. Crown soda bottles used by Illinois firms (Galesburg to Lincoln).

FIGURE 81. Crown soda bottles used by Illinois firms (Litchfield to Quincy).

Figure 82. Crown soda bottles used by Illinois firms (Quincy to Wyoming).

Appendix

A CHRONOLOGY OF SOME SOFT DRINK BRANDS IN AMERICA*

1871

Lemon's Superior Sparkling Ginger Ale, the first soft drink trademark registration.

1876

Charles E. Hires of Philadelphia began the manufacture of his root beer extract.

1880

James Vernor of Detroit developed Vernor's Ginger Ale.

1881

The Clicquot Club Company of Millis, Massachusetts, began producing ginger ale and other beverages.

1883

White Rock mineral waters were first bottled at Waukesha, Wisconsin.

1885

The Moxie Nerve Food Company of Boston began bottling and shipping their new soft drink, Moxie.

1886

Coca-Cola is produced and marketed by its originator, John S. Pemberton of Atlanta, Georgia.

1888

The Artesian Manufacturing and Bottling Company of Waco, Texas, bottled the first Dr. Pepper.
New carbonated beverages bottled by John J. McLaughlin of Toronto, Canada, became known as Canada Dry.

1889

Coca-Cola adopts the slogan, "Delicious and Refreshing."

1890

Dr. Swetts Root Beer first appears as a new soft drink.

1893

Hires Root Beer is bottled for the first time by the Crystal Bottling Company of Philadelphia, Pennsylvania.

1894

The first bottling of Coca-Cola by Joseph A. Biedenharn of Vicksburg, Mississippi.

1896

Brad's Drink is developed by Caleb D. Bradham of New Bern, North Carolina. In 1901 it became known as Pepsi-Cola.

1905

Iron Brew and Iron Beer appear on the market as new soft drinks.

1906

New soft drinks from southern United States include Buffalo Rock Ginger Ale and Barq's Root Beer.

1912

Claude A. Hatcher of Columbus, Georgia, organizes the Chero Cola Company.

1913

A series of new drinks including Orange Smile, Cherry Blossoms, Orange Julep, Grape Julep, Cherry Julep, and Cherry Smash are introduced and soon become popular.

1915

Delaware Punch is introduced as one of several new cloudy-type orange drinks.

1916

Orange Crush and Whistle are introduced as new franchised bottle drinks.

1920

The Howdy Beverage Company is organized by C. L. Grigg in St. Louis.

1921

The Nu Grape Company of America is organized in Atlanta under the management of O. R. Randall.

*** After Riley, 1958**

1922

Various grape drinks including Nu Grape, Good Grape, Grapico, Bluebird, Nuicy, Brandimist and Brandywine become popular.

Orange Kist and Ginger-Mint Julep appear as new soft drinks.

1924

The Chero Cola Company introduces a new line of flavored drinks called Nehi.
Nesbitt's Orange is another of several new franchised drinks.
The Dr. Pepper Company is organized in Dallas, Texas.

1928

C. L. Grigg, founder of the Howdy Company of St. Louis, introduces Seven-Up.

1929

Coca-Cola first uses the slogan, "The Pause That Refreshes."

1930

The chocolate-drink era begins, and the new brands include Chocolate Soldier, Mavis, Coc-au-lait, Brownie, Chocolate Mello, Kayo, and Kremko.

1933

Some newly introduced brands include Ju'Cy Orange, Par-t-Pak, Upper 10, Mission Orange.

1935

The Nehi Corporation introduces Royal Crown Cola.

1936

Newly introduced drinks include Double Cola, Try Me, Big Bill, Big Boy, 2-Way, Pick Upper.

1937

New root beer brands include Twang, Howell's, Dad's Old Fashioned and Tripple XXX.

1938

Squirt, Fruit Bowl, Tru-Ade, Suncrest Orange, Sun Spot and Lucky Club are introduced.

BOTTLE AND BOTTLING INDUSTRY LITERATURE

The earliest literature in the soda water and carbonated beverage trade consisted largely of books on practical and theoretical chemistry. The major advances in the industry were initially made by chemists and others of a scientific bent; consequently, little material of a strictly popular nature appeared in print. It is said that John Matthews (of the firm by the same name) issued a catalog as early as 1860. Other catalogs by this firm were to follow in subsequent years, and soon several companies and individuals, including the famed James W. Tufts (*American Soda Book*), were publishing occasional volumes. It was not until the 1880s that any major texts were printed in America which could be utilized as a guide by bottlers. Although several were subsequently printed during the 1880s, the *Treatise On Beverages—The Complete Practical Bottler* by Charles Herman Sulz was the most complete and reliable work of its era.

The first trade journals were also started during this period. *The American Carbonator and American Bottler*, started in 1881 in New York City, was devoted to the general bottling trade and soda water counter. The name of this journal was changed to *The American Bottler* in 1906, and in May 1924 it ceased publication. In 1882 a second major trade journal was organized, the *National Bottlers' Gazette*, and it has enjoyed the distinction of being the longest-lived periodical in the carbonated beverage field.

Several other journals were published during the latter years of the 19th century; however, most had a short life or a heavy emphasis on other aspects of the bottling trade. All, however, offer interesting commentaries on the early days of the beverage trade and are of interest to historians and collectors alike. Copies of these old journals, books, and catalogs are only rarely brought to light. Information found in the following references should prove both interesting and enlightening to the serious student or collector of soda and mineral water bottles and related accessories.

Anonymous. 1965. *The Pepsi-Cola Story.* Pepsi-Cola Co., N.Y., N.Y.

Antiques Research Publications. 1968. *Soda Fountain Supplies and Furnishings of the Early 1900s.* Chattanooga, Tenn.

Cope, Jim. 1971. *Soda Water Advertising.* Pub. by author, Orange, Texas.

Fitch, William E. 1927. *Mineral Waters of the United States.* Lea & Febiger, Philadelphia & New York.

Fountain, John C., and Donald Colcleaser. 1968. *Dictionary of Soda & Mineral Water Bottles.* Ole Empty Bottle House Pub. Co., (Pub. by authors), Amador City, Calif.

Garrett, Franklin. 1968. *Chronological History of the Coca-Cola Company 1886-1967.* Coca-Cola Co., Atlanta, Ga.

Lincoln, Gerald D. 1970. *Antique Blob Top Bottles, Central and Southern New England.* Pub. by author, Marlboro, Mass.

Morrison, Joseph L. 1962. The Soda Fountain. *Amer. Heritage*, Vol. 8, No. 5.

Munsey, Cecil. 1970. *An Illustrated Guide to Collecting Bottles.* Hawthorn Books, Inc., N.Y.

Riley, John J. 1958. *A History of the American Soft Drink Industry 1807-1957.* Amer. Bottlers of Carbonated Beverages, Wash., D.C.

Schmeiser, Alan. 1968. *Have Bottles Will Pop.* Michalan Press, (Pub. by author), Dixon, Calif.

Schmeiser, Alan. 1970. *More Pop.* Michalan Press, (Pub. by author), Dixon, Calif.

Toulouse, Julian H. 1971. *Bottle Makers and Their Marks.* Thomas Nelson Inc., N.Y.

Tufts, James W. 1885(?) (reprint 1969). *The Manufacture and Bottling of Carbonated Beverages.* Frontier Book Co., Fort Davis, Texas.

Tufts, James W. 1895(?) *Sodawater Apparatus Book of Directions.* Geo. H. Ellis, Printer, Boston, Mass.